CONTENTS

INTRODUCTION

When this volume was last revised, building was controlled by The Building Regulations 1976 which contained regulations and deemed to satisfy provisions for building together with schedules containing technical details of materials and construction that met the provisions in the Regulations.

Since then, in a new format, The Building Regulations 1985 has come into operation. These, the current regulations, which set out minimum functional requirements in eleven parts, contain no technical details as to how the requirements may be met. The stated purpose in drafting the Regulations in this way is to allow flexibility in their interpretation.

Published with The Building Regulations 1985 are eleven Approved Documents as practical guidance to meeting the requirements of the Regulations. Although there is no obligation to adopt any solution detailed in the Approved Documents it is stated that 'If a contravention of a requirement is alleged then, if you have followed the guidance in the Approved Document, that will be evidence tending to show that you have complied with the Regulations'.

The consequence of this is that the technical guidance in the Approved Documents will generally be accepted as the requirements of the Regulations, to avoid the lengthy and time consuming work of referring to a bewildering array of British Standards and Board of Agrément Certificates in support of a solution not contained in the Approved Documents.

The changes in the Building Regulations between 1976 and 1985 are mostly in format, the latter having a more readable and attractive layout. In fact, the changes are more in presentation than content, as the Approved Documents in the main replace the Schedules contained in the 1976 Regulations, with updating of references and the addition of some diagrams. It appears that the authorities toyed with the idea of giving more freedom of choice in the form of building and then withdrew to the safety of detailed regulation through the device of Approved Documents that might be ignored at your peril.

In this revision the content of the four chapters, foundations, walls, floors and roofs had been revised, updated and rearranged. In each chapter is a general explanation of the functional requirements relevant to the subject matter followed by details of materials and construction with diagrams, to meet the requirements of the Building Regulations for small buildings, such as houses, constructed in the main with traditional materials. Details of the 1989 revision of the practical guidance contained in Approved Documents F and L, Ventilation and Resistance to the passage of heat, are included.

AUTHOR'S NOTE

For linear measure all measurements are shown in either metres or millimetres. A decimal point is used to distinguish metres and millimetres, the figures to the left of the decimal point being metres and those to the right millimetres. To save needless repetition, the abbreviations 'm' and 'mm' are not used, with one exception. The exception to this system is where there are at present only metric equivalents in decimal fractions of a millimetre. Here the decimal point is used to distinguish millimetres from fractions of a millimetre, the figures to the left of the decimal point being millimetres and those to the right being fractions of a millimetre. In such cases the abbreviation 'mm' will follow the figures e.g. 203.2 mm.

R. Barry

CHAPTER ONE

FOUNDATIONS AND OVERSITE CONCRETE

The foundations of buildings bear on and transmit loads to the ground. The foundation is that part of walls, piers and columns in direct contact with and transmitting loads to the ground. In practice, the concrete base of walls, piers and columns is described as the foundation and that part of a wall, pier or column below ground and below the horizontal d.p.c. as the foundation wall, pier or column. The principal foundation types are strip, pad, raft and pile foundations as illustrated in Fig.1.

It will be seen that a strip foundation is a continuous strip of concrete under walls, a pad an isolated base under piers and columns, a raft a continuous base under the whole of the building, and a pile a concrete column or pillar cast in or driven into the ground to support a concrete base or ground beam.

Ground is the general term for the earth's surface, which varies in composition within the two main groups, rocks and soils. Rocks include the hard, rigid, strongly-cemented geological deposits such as granite, sandstone and limestone, and soils the comparatively soft, loose, uncemented geological deposits such as gravel, sand and clay. Unlike rocks, soils compact under the compression of the foundation loads of buildings.

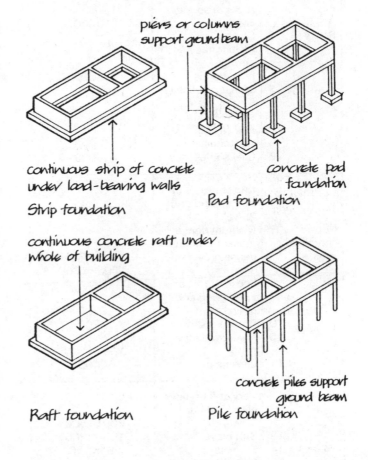

Fig. 1

FUNCTIONAL REQUIREMENTS

The functional requirements of foundations and oversite concrete are:

Strength and stability
Resistance to ground moisture
Resistance to the passage of heat

Strength and stability

The requirements from Part A of Schedule 1 to The Building Regulations 1985 are, as regards 'Loading', that 'The building shall be so constructed that the combined, dead, imposed and wind loads are sustained and transmitted to the ground safely and without causing such deflection or deformation of any part of the building, or such movement of the ground, as will impair the stability of any part of another building' and as regards 'Ground movement' that 'The building shall be so constructed that movements of the subsoil caused by swelling, shrinkage or freezing will not impair the stability of any part of the building'.

A foundation should be designed to transmit the loads of the building to the ground so that there is, at most, only a limited settlement of the building into the ground. A building whose foundation is on sound rock will suffer no measurable settlement whereas a building on soil will suffer settlement into the ground by the compression of the soil under the

1

Table 1. Typical Bearing Capacities

Group		Types of rocks and soils	Bearing capacity (kN/m²)
I	1	Strong igneous and gneissic rocks in sound condition	10 000
Rocks	2	Strong limestones and strong sandstones	4 000
	3	Schists and slates	3 000
	4	Strong shales, strong mudstones and strong siltstones	2 000
	5	Clay shales	1 000
II	6	Dense gravel or dense sand and gravel	>600
Non-cohesive soils	7	Medium dense gravel or medium dense gravel and sand	<200 to 600
	8	Loose gravel or loose sand and gravel	<200
	9	Compact sand	>300
	10	Medium dense sand	100 to 300
	11	Loose sand	<100
III	12	Very stiff boulder clays and hard clays	300 to 600
	13	Stiff clays	150 to 300
Cohesive soils	14	Firm clays	75 to 150
	15	Soft clays and silts	75
IV	16	Peat and organic soils	Foundations carried down through peat to a reliable bearing stratum
V	17	Made ground or fill	Should be investigated with extreme care

Based on BS 8004:1986

foundation loads. Foundations should be designed so that settlement into the ground is limited and uniform under the whole of the building. Some settlement of a building on a soil foundation is inevitable as the increasing loads on the foundation, as the building is erected, compress the soil. This settlement must be limited to avoid damage to service pipes and drains connected to the building. Bearing capacities for various rocks and soils are assumed and these capacities should not be exceeded in the design of the foundation to limit settlement. These bearing capacities are set out in table 1 on page 2.

In theory, if the foundation soil were uniform and foundation bearing pressure were limited, the building would settle into the ground uniformly as the building was erected, and to a limited extent, and there would be no possibility of damage to the building or its connected services or drains. In practice there are various possible ground movements under the foundation of a building that may cause one part of the foundation to settle at a different rate and to a different extent than another part of the foundation. This different or differential settlement must be limited to avoid damage to the superstructure of the building. Some structural forms can accommodate differential or relative foundation movement without damage more than others. A brick wall can accommodate limited differential movement of the foundation or the structure by slight movement of the small brick units and mortar joints, without affecting the function of the wall, whereas a rigid framed structure with rigid panels cannot to the same extent. Foundations are designed to limit differential settlement, the degree to which this limitation has to be controlled or accommodated in the structure depends on the nature of the structure supported by the foundations.

ROCKS AND SOILS

Rocks are divided into three broad groups: sedimentary, metamorphic and igneous according to their geological formation. Sedimentary rocks are those formed gradually over thousand of years from particles of calcium carbonate or sand deposited by settlement in bodies of water and gradually compacted into rocks such as sandstone and limestone. Metamorphic rocks are those changed from igneous, sedimentary, or from earth into metamorphic by pressure or heat or both, such as slates and schists. Igneous rocks are those formed by the fusion of minerals under great heat and pressure such as granite, dolerite and basalt. A classification of rocks is set out in table 2.

Table 2. Classification of Rocks

Group	Rock type	
Sedimentary	Sandstones (including conglomerates)	Siliceous Calcareous Ferruginous Argillaceous
	Some hard shales and tuffs	
	Limestones	Massively bedded (including chalk) Thinly bedded
Metamorphic	Some hard shales Slates Schists Gneisses	
Igneous	Granite Dolerite Basalt	

Soils

Top soil: The surface of much of the land in this country is covered with a layer of top soil or vegetable soil to a depth of about 100–300. Top soil is composed of loose soil, growing plant life and the accumulation of decaying vegetation. As it is very soft it is unsatisfactory as a foundation and it is stripped from the site of buildings.

Subsoil: The soil below the top soil is termed subsoil.

Soils are grouped by reference to the size and nature of the particles and the density of the particular soil. The three broad groups are coarse grained non-cohesive, fine grained cohesive and organic. The nature and behaviour under load of the soils in each group are similar. Organic soils, such as peat, are not generally suitable as a foundation for buildings.

Coarse grained non-cohesive soils such as sands and gravels consist of the coarser largely siliceous un-altered products of rock weathering. They have no plasticity and tend to lack cohesion especially when dry. Under pressure from the loads on foundations the soils in this group compress and consolidate rapidly by some rearrangement of the particles and the expulsion of water. A foundation on coarse grained non-cohesive soils settles rapidly by consolidation of the soil, as the building is erected, so that there is no further settlement once the building is completed. Some sand or sandy soils in waterlogged conditions may expand in frost (frost heave) at and for some distance below the surface. In exposed situations it is practice to carry the foundations down to a depth of 450, which is below the frost line, to avoid ground movement in soils liable to frost heave under and around unheated buildings.

Gravel. A natural coarse grained deposit of rock fragments and finer sand. Many of the particles are larger than 2.

Sand. A natural sediment of granular, mainly siliceous products of rock weathering. Particles are smaller than 2, are visible to the naked eye and the smallest size is 0.06 mm. Sand is gritty, has no real plasticity and can be easily powdered by hand when dry.

Fine grained cohesive soils such as clay, are a natural deposit of the finest siliceous and aluminous products of rock weathering. Clay is smooth and greasy to the touch, shows high plasticity, dries slowly and shrinks appreciably on drying. Under pressure of the load on foundations clay soils are very gradually compressed by the expulsion of water through the very many, fine capillary paths so that buildings settle gradually during building work and this settlement may continue for some years after the building is completed.

Firm shrinkable clays suffer appreciable vertical and horizontal shrinkage on drying and expansion on wetting. Shrinkage and expansion of firm clays under grass extends to about 1 m below the surface in Great Britain and up to depths of 4 m or more below large trees. The extent of volume changes, particularly in firm clay soils, depends on seasonal variations and the proximity of trees and shrubs. The greater the seasonal variation, the greater the volume change. The more vigorous the growth of shrubs and trees in firm clay soils the greater the depth below surface the volume change will occur.

As a rough guide it is recommended that buildings on shallow foundations should not be closer to single trees than the height of the tree at maturity and one-and-a-half times the height at maturity of groups of trees, to reduce the risk of damage to buildings by seasonal volume changes in clay subsoils.

When shrubs and trees are removed to clear a site for building on firm clay subsoils there will, for some years after clearance, be ground recovery as the clay

Table 3. Minimum Width of Strip Foundations

Type of subsoil	Condition of subsoil	Field test applicable	Total load of load bearing walling not more than (kN/linear m)					
			20	30	40	50	60	70
			Minimum width of strip foundation (mm)					
I Rock	Not inferior to sandstone, limestone or firm chalk	Requires at least a pneumatic or other mechanically operated pick for excavation	In each case equal to the width of wall					
II Gravel Sand	Compact Compact	Requires pick for excavation. Wooden peg 50 mm square in cross-section hard to drive beyond 150 mm	250	300	400	500	600	650
III Clay Sandy clay	Stiff Stiff	Cannot be moulded with the fingers and requires a pick or pneumatic or other mechanically operated spade for its removal	250	300	400	500	600	650
IV Clay Sandy clay	Firm Firm	Can be moulded by substantial pressure with the fingers and can be excavated with graft or spade	300	350	450	600	750	850
V Sand Silty sand Clayey sand	Loose Loose Loose	Can be excavated with a spade. Wooden peg 50 mm square in cross-section can be easily driven	400	600				
VI Silt Clay Sandy clay Silty clay	Soft Soft Soft Soft	Fairly easily moulded in the fingers and readily excavated	450	650	Note: in relation to types V, VI and VII, foundations do not fall within the provisions of this section if the total load exceeds 30 kN/m			
VII Silt Clay Sandy clay Silty clay	Very soft Very soft Very soft Very soft	Natural sample in winter conditions exudes between fingers when squeezed in fist	600	850				

Taken from Part E of Approved Document A of The Building Regulations 1985

gradually recovers moisture previously withdrawn by the shrubs and trees. This gradual recovery of water by the clay and consequent expansion may take several years. The depth at which the recovery and expansion is appreciable will be roughly proportional to the height of the trees and shrubs removed, and the design and depth of foundations of buildings must allow for this gradual expansion to limit damage by differential settlement. Similarly if vigorous shrub or tree growth is stopped by removal, or started by planting, near to a building on a firm clay subsoil it is most likely that gradual expansion or contraction of the soil will cause damage to the building by differential movement.

SITE INVESTIGATION AND EXPLORATION

To select a foundation from tables or to design a foundation it is necessary to calculate the loads on the foundation and determine the nature of the subsoil, its bearing capacity, its likely behaviour under seasonal and ground water level changes and the possibility of ground movement. Where the nature of the subsoil is known from geological surveys, adjacent building work or trial pits or borings and the loads on foundations are small, as for single domestic buildings, it is generally sufficient to excavate for foundations and confirm, from the exposed subsoil in the trenches, that the soil is as anticipated.

Under strip and pad foundations there is a significant pressure on the subsoil below the foundation to a depth and breadth of about one-and-a half times the width of the foundation. If there were, in this area below the foundation, a soil with a bearing capacity less than that below the foundation, then appreciable settlement of the foundation might occur and damage the building. It is important, therefore, to know or ascertain the nature of the subsoil both at the level of the foundation and for some depth below.

The required width of foundation for buildings of up to three storeys may be determined from table 3, on page 4 when the loads and the nature of the subsoil are known. The depth of the foundation below ground will depend on the nature of the subsoil and the need to limit differential settlement due to ground movements.

Where the nature of the subsoil is uncertain or there is a possibility of ground movement or a need to confirm information on subsoils, it is wise to explore the subsoil over the whole of the site of the building.

As a first step it is usual to collect information on soil and subsoil conditions from the County and Local Authority whose local knowledge from maps, geological surveys, aerial photography and works for buildings and services adjacent to the site may in itself give an adequate guide to subsoil conditions. In addition geological maps from the British Geological Survey, information from local geological societies, Ordnance Survey maps, mining and river and coastal information may be useful.

A visit to the site and its surroundings should always be made to record everything relevant from a careful examination of the nature of the subsoil, vegetation, evidence of marshy ground, signs of ground water and flooding, irregularities in topography, ground erosion and ditches and flat ground near streams and rivers where there may be soft alluvial soil. A record should be made of the foundations of old buildings on the site and cracks and other signs of movement in adjacent buildings as evidence of ground movement.

To make an examination of the subsoil on a building site, trial pits or boreholes are excavated. Trial pits are usually excavated by machine or hand to depths of 2–4 m and at least the anticipated depth of the foundations. The nature of the subsoil is determined by examination of the sides of the excavations. Boreholes are drilled by hand auger or by machine to withdraw samples of soil for examination. Details of the subsoil should include soil type, consistency or strength, soil structure, moisture conditions and the presence of roots at all depths. From the nature of the subsoil the bearing capacity, seasonal volume changes and other possible ground movements are assumed. To determine the nature of the subsoil below the foundation level it is either necessary to excavate trial pits some depth below the foundation or to bore in the base of the trial hole to withdraw samples. Whichever system is adopted will depend on economy and the nature of the subsoil. Trial pits or boreholes should be sufficient in number to determine the nature of the subsoil over and around the site of the building and should be at most say 30 m apart.

Ground movements that may cause settlement are:

(a) compression of the soil by the load of the building
(b) seasonal volume changes in the soil
(c) ground recovery due to the felling of trees
(d) mass movement in unstable areas such as made-up ground and mining areas where there may be considerable settlement
(e) ground made unstable by adjacent excavations or by dewatering, for example, due to an adjacent road cutting.

It is to anticipate and accommodate these movements that site investigation and exploration is carried out. For further details of site investigation and exploration see Volume 4.

Compression of the soil by the load of the building

Differential settlement may occur where there is a significant difference in the loads on adjacent parts of the foundation or where there is a difference in the bearing capacity of the subsoil below the building. From calculations of the anticipated loads on the foundations and from an investigation of the subsoil, differential settlement may be limited in the design of the foundation.

Seasonal volume changes in the soil

Seasonal changes and the withdrawal of moisture by deep-rooted vegetation can cause considerable volume changes in shrinkable clay soils especially after periods of low rainfall. These volume changes can cause severe damage to low rise buildings with foundations near the surface.

The extent of volume changes in shrinkable clay soils and the depth below the surface at which these changes can affect foundations depends largely on whether the building is sited on open ground away from deep-rooted vegetation or close to past, present or future growth of deep-rooted vegetation. A foundation depth of at least 0.9 m on shrinkable clay soils on open ground away from deep-rooted vegetation has been recommended by the Building Research Station for more than thirty years and recent experience, such as the severe drought of 1976, has affirmed the wisdom of this recommendation. There appears to be little if any evidence of shrinkage damage to buildings with foundations at or below the recommended depth during periods of drought.

At the recommended depth of at least 0.9 m it is not generally economic to use the traditional strip foundation and hence the narrow strip or trench fill foundation has been used (Fig. 5). A narrow trench 400 wide is excavated by machine and filled with concrete to just below the surface. If the concrete is placed immediately after the excavation there is no need to support the sides of the trench in stiff clays, the sides of the trench will not be washed away by rain and the exposed clay will not suffer volume change by exposure. A narrow strip or trench fill foundation is appreciably cheaper than a traditional strip foundation with foundation walls at these depths.

The foundations of buildings sited adjacent to past, present or future deep-rooted vegetation can be affected at a considerable depth below the surface by the gain or removal of ground moisture and consequent expansion or shrinkage. Appreciable expansion, following the removal of deep-rooted vegetation, may continue for some years as the subsoil gains moisture. Significant seasonal volume change, due to deep-rooted vegetation, will be pronounced during periods of drought and heavy continuous rainfall. The vigorous growth of newly planted deep-rooted vegetation adjacent to buildings may cause continuous shrinkage in clay soils for some years. The most economical and effective foundation for low rise buildings on shrinkable clays close to deep-rooted vegetation is a system of short-bored piles and ground beams (Fig. 13). The piles should be taken down to a depth below which vegetation roots will not cause significant volume changes in the subsoil. Single deep-rooted vegetation such as shrubs and trees as close as their mature height to buildings, and groups of shrubs and trees one-and-a-half times their mature height to buildings can affect foundations on shrinkable clay subsoils.

Frost heave

Where the water table is high, that is near the surface, soils, such as silts, chalk, fine gritty sands and some lean clays, near the surface may expand when frozen. This expansion, or frost heave, is due to crystals of ice forming and expanding in the soil and so causing frost heave. In this country, ground water near the surface rarely freezes at depths of more than 0.5 m but in exposed positions on open ground during frost it may freeze up to a depth of 1 m. Even in exposed positions during severe frost it is most unlikely that ground water under and adjacent to the foundations of heated buildings will freeze because of the heat stored in the ground under and around the building. There is, therefore, no need to consider the possibility of ground movement due to frost heave under and around heated buildings.

For unheated buildings and heated buildings with insulated ground floors, a foundation depth of 450 is generally sufficient against the possibility of damage by ground movement due to frost heave.

Ground recovery due to the felling of trees

When deep-rooted shrubs and trees are cleared for building on shrinkable clay subsoils there will be a gradual recovery of ground moisture to the soil for some years after the clearance and a consequent gradual expansion of the soil for some depth below the surface. It is necessary, therefore, to take foundations down to a depth at which there will be no significant ground movement by the use, for example, of short-bored piles.

Mass movement in unstable areas such as made-up ground and mining areas

The surface of ground that has been raised by filling or tipping soil, waste or refuse is described as made ground or made-up ground. These filled or fill sites on made-up ground should be avoided for building because of the extreme variability of the nature of the fill and the difficulty of predicting the degree of compaction under load and the variability of compaction and settlement. Where buildings are to be sited on made-up ground a thorough investigation of the nature of the fill must be carried out to anticipate the likelihood of differential settlement. Single or semi-detached houses on fill should have a raft foundation with an edge beam (see Fig. 11). The construction of terraced houses on fill should be avoided because of the likelihood of differential settlement causing damage by cracking. Larger buildings should be supported by a pile foundation carried down to a firm base. In areas liable to mining subsidence it is wise to seek the advice of the National Coal Board on likely ground movement and recommended foundations. For two storey houses the raft foundation, illustrated in Fig. 10, is often recommended, the thin reinforced concrete raft being laid on a bed of fine granular material to accommodate some ground movement independent of the slab. The CLASP system, illustrated on page 38 of Volume 4, was developed specifically for school buildings on areas liable to mining subsidence.

Concrete in sulphate-bearing soils

Water-soluble sulphates, particularly in clay soils, can combine with the set cement in concrete and cause expansion, cracking and disintegration of concrete. The conditions most favourable to this action are concrete mixes of cements, such as ordinary Portland cement, which combine readily with the sulphates in water-saturated sulphate-bearing soils at or below the water table. The continuous migration of sulphates through ground water to concrete may cause the continuing combination of sulphates with cement and the gradual expansion, cracking and ultimate disintegration of the concrete.

Concrete which is permanently above the water table is unlikely to be attacked by sulphates.

Large sections of thoroughly compacted, dense, impermeable concrete are less subject to attack by sulphates than poorly compacted, pervious concrete into which the sulphates will more readily penetrate in solution in ground water.

Where concrete is placed in sulphate-bearing soils it should consist of one of the sulphate-resisting cements such as sulphate-resisting Portland cement and be fully compacted.

Chemical compounds in fill material have caused disintegration of foundation slabs and brickwork. An impermeable membrane between the fill material, likely to cause damage, and the foundation and slabs will reduce the chance of damage.

FOUNDATION TYPES

The four principal types of foundation are strip, pad, raft and pile foundations as illustrated in Fig. 1.

Strip foundations

Strip foundations consist of a continuous strip, usually of concrete formed centrally under load bearing walls. The continuous strip serves as a level base on which the wall is built and is of such a width as is necessary to spread the load on the foundations to an area of subsoil capable of supporting the load without undue compaction. Concrete is the material principally used today for foundations as it can readily be placed, spread and levelled in foundation trenches, to provide a base for walls, and it develops adequate compressive strength as it hardens to support the load on foundations. Before Portland cement was manufactured, strip foundations of brick were common, the brick foundation being built directly off firm subsoil or built on a bed of natural stones. The brick foundation was built in steps as illustrated in Fig. 2. A brick foundation is little used today in this country.

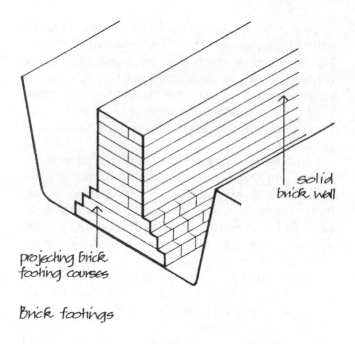

solid brick wall

projecting brick footing courses

Brick footings

Fig. 2

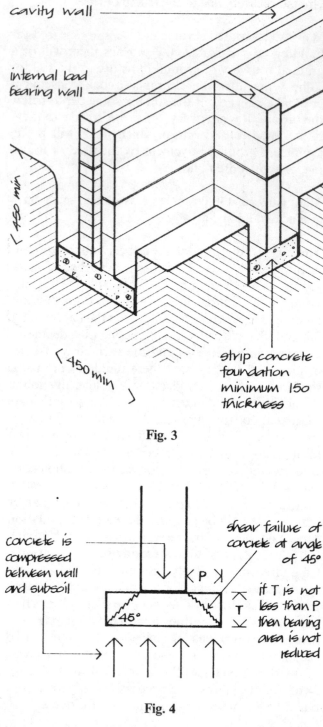

cavity wall

internal load bearing wall

450 min

450 min

strip concrete foundation minimum 150 thickness

Fig. 3

concrete is compressed between wall and subsoil

shear failure of concrete at angle of 45°

45°

< P >

T

if T is not less than P then bearing area is not reduced

Fig. 4

The width of a concrete strip foundation depends on the bearing capacity of the subsoil and the load on the foundations. The greater the bearing capacity of the subsoil the less the width of the foundation for the same load. The minimum width of strip foundations is set out in table 3, taken from Approved Document A to The Building Regulations 1985. The minimum thickness of concrete is 150 or the projection of the strip each side of the wall, whichever is the greater as shown in Fig. 3. which is an illustration of the strip foundation for a cavity wall on soft clay.

The thickness of the concrete should be not less than the projection of the strip each side of the wall, as if there were a failure of the concrete by shear, the 45° angle of shear would not reduce the bearing of the base on the subsoil as illustrated in Fig. 4. The practical minimum depth of a strip foundation is usually 450 to allow for the removal of top soil and variations in ground level and the usual width is 450 to provide space in the trench to lay foundation brickwork.

Narrow strip (trench fill) foundations: Where the bearing capacity of the subsoil and the loads on the foundations require the strip to be little wider than the thickness of the wall yet the nature of the subsoil such as clay may require a depth of up to 0.9 m, it is usual to excavate foundation trenches and fill them with concrete up to a level just below the finished ground level as illustrated in Fig. 5. It is cheaper to fill the trenches with concrete than excavate a wider trench to provide room for building the wall below ground. These narrow trench fill foundations are commonly excavated by machine.

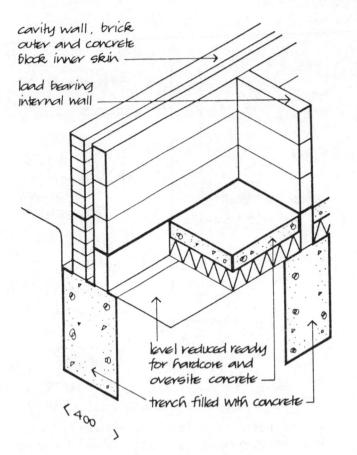

cavity wall, brick
outer and concrete
block inner skin

load bearing
internal wall

level reduced ready
for hardcore and
oversite concrete

trench filled with concrete

< 400 >

Narrow trench fill foundation

Fig. 5

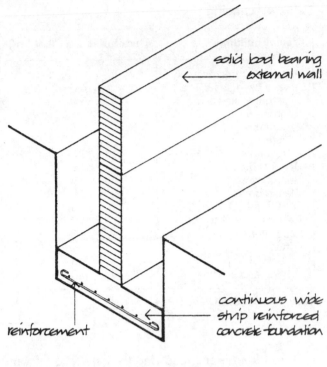

solid load bearing
external wall

reinforcement

continuous wide
strip reinforced
concrete foundation

Wide strip foundation

Fig. 6

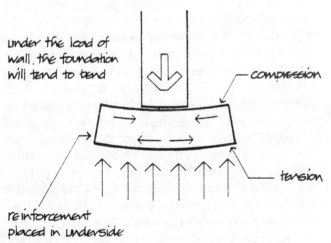

under the load of
wall, the foundation
will tend to bend

compression

tension

reinforcement
placed in underside

Fig. 7

Wide strip foundations: Where the loads on foundations are considerable in relation to the bearing capacity of the subsoil, it is necessary to use a wide strip foundation to spread the load. Rather than make the thickness of the strip of concrete equal to the projection of the strip each side of the wall, it is generally cheaper to reinforce the concrete strip to reduce its thickness as illustrated in Fig. 6. Concrete is strong in compression and weak in tension and the steel reinforcement is placed in the concrete strip where tensile stress is greatest. Under load the concrete strip will tend to bend as shown in Fig. 7 and the reinforcement is therefore cast into the underside of the strip.

Loads on foundation

The load on the foundation is the accumulation of the dead loads, imposed loads and wind load.

The dead load is the force due to the static weight of all walls, partitions, floors, roofs and finishes and all other permanent construction. The unit weights of building materials is set out in BS 648.

Imposed load is the load assumed to be produced by the intended use of the building, including distributed, concentrated, impact, inertia and snow loads but excluding wind loads. Some common distributed loads are set out in table 4.

9

Table 4. Loading

Use of building	Distributed load (kN/m²)
Assembly buildings	
with fixed seats	4.0
without fixed seats	5.0
Bedrooms	
domestic	1.5
hotels	2.0
hospitals	2.0
Boiler rooms	7.5
Bungalows	1.5
Classrooms	3.0
Dwellings	1.5
Factories, workshops	5.0
Laboratories	3.0
Offices	
general	2.5
filing and storage	5.0

Taken from BS 6399:Part 1:1984

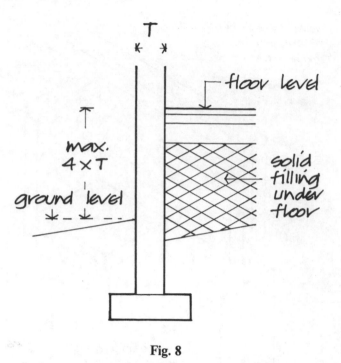

Fig. 8

Wind load is the load due to the effects of wind pressure or suction.

In the calculation of loads on foundations for load bearing walls it is usual to calculate the load per metre length of wall by calculating the sum of the dead, imposed and wind loads per metre of wall and from this to determine the required width of strip foundation from the type of soil on which the foundation is to bear. In the table from Approved Document A to The Building Regulations 1985 (table 3, on page 4) the required width of foundations for given loads and soils are shown. In small buildings, such as houses, where the loads are small it is usual to assume a uniform spread of loads along walls so that a calculation of load on a metre run of wall between windows may be taken to select the width of foundation for the whole wall.

Where there are differences of level of the ground and solid construction between one side of a wall and the other, the difference of level must not be more than four times the thickness of the wall as illustrated in Fig. 8, to limit the lateral force on the wall from the higher solid filling under the floor.

Pad foundations

It is sometimes economical to construct a foundation of isolated piers or columns of brick or concrete supporting reinforced concrete ground beams, in turn supporting walls, rather than excavating deep trenches and raising walls off strip foundations, some depth below ground. Where the subsoil has poor bearing capacity for some depth below the surface, as for example where the ground has been made up, it is often economical to use a foundation of piers on pad foundations, as illustrated in Fig. 9. The isolated concrete pad foundations are spread in the base of excavations, on which piers or columns of brick or concrete are raised to ground level to support reinforced concrete ground beams off which the walls are raised. The spread of the pad foundation is determined by the loads on it and the bearing capacity of the subsoil, and the thickness of the concrete is either at least equal to the projection of the pad each side of the pier or the pad foundation is reinforced. The spacing of the piers or columns is determined by the most economical construction.

Raft foundations

Raft foundations consist of a raft of reinforced concrete under the whole of the building designed to transmit the load of the building to the subsoil below the raft. Raft foundations are used for buildings on compressible ground such as very soft clays, alluvial deposits and compressible fill material where strip foundations would not provide a stable foundation.

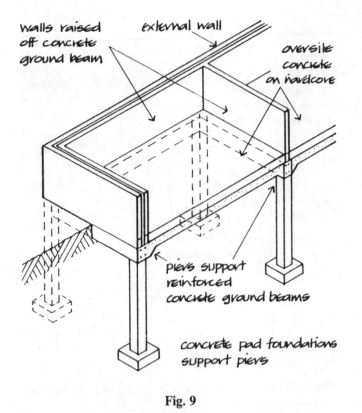

walls raised off concrete ground beam

external wall

oversite concrete on hardcore

piers support reinforced concrete ground beams

concrete pad foundations support piers

Fig. 9

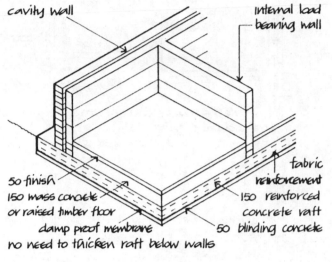

cavity wall

internal load bearing wall

50 finish
150 mass concrete or raised timber floor
damp proof membrane
no need to thicken raft below walls

fabric reinforcement

150 reinforced concrete raft

50 blinding concrete

Fig. 10

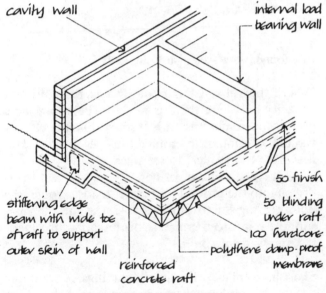

cavity wall

internal load bearing wall

stiffening edge beam with wide toe of raft to support outer skin of wall

reinforced concrete raft

50 finish

50 blinding under raft

100 hardcore

polythene damp proof membrane

Fig. 11

The two types of raft commonly used are the flat raft and the wide toe raft foundation as illustrated in Figs 10 and 11. The flat slab raft is cast on a bed of blinding concrete and a moisture-proof membrane to prevent damp rising through the slab. As will be seen from Fig. 10 the slab is reinforced top and bottom and is of uniform thickness. Where the ground has a reasonable bearing capacity the raft may not need to be reinforced. For small buildings, such as two storey houses, there is no need to thicken the raft either under the external or internal load bearing walls.

A flat slab recommended for building in areas subject to mining subsidence is similar to the flat slab, but cast on a bed of fine granular material 150 thick so that the raft is not keyed to the ground and is therefore unaffected by horizontal ground strains.

Where the ground has poor compressibility the wide toe raft is recommended, the stiffening edge beam being designed as a toe to support the outer skin of brickwork, so that the raft is not visible (Fig. 11). The raft is strengthened with a reinforced concrete edge beam under external walls and is thickened under internal load bearing walls. There is no structural advantage in extending the edge beam

down into the ground as though it were a strip foundation nor is there any need to extend the edge beam into the ground below the frost line.

On sloping sites a raft foundation can be used either on compacted levelled fill above the natural ground line as illustrated in Fig. 12, or in steps cut into the natural ground as illustrated in Fig. 12 or as cut and fill as illustrated in Fig. 16.

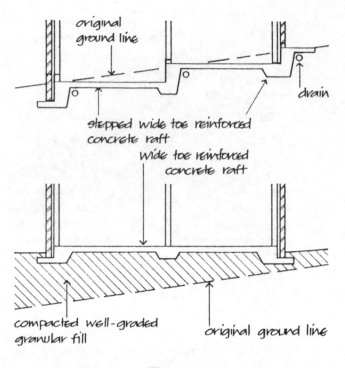

Fig. 12

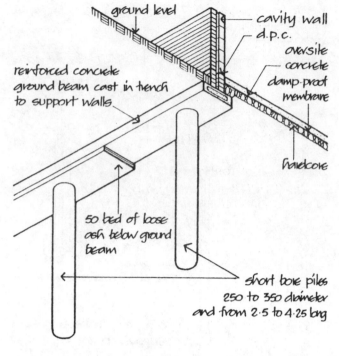

Short Bored pile foundation [worms eye view]

Fig. 13

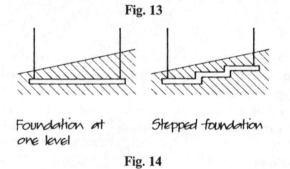

Foundation at one level   Stepped foundation

Fig. 14

Pile foundations (see also Volume 4)

Where the subsoil has poor or uncertain bearing capacity or where there is likely to be appreciable ground movement as with firm, shrinkable clay or where the foundation should be deeper than say 2 m, it is often economical to use piles.

A pile is a column of concrete either cast in or driven into the ground to transfer loads through the poor bearing soil to a more stable stratum. The piles support reinforced concrete beams off which load-bearing walls are built.

Short-bored piles: For small buildings, for example on shrinkable clays where adjacent trees or the felling of trees makes for appreciable volume change in the subsoil for some depth, it is often wise and economical to use a system of short-bored piles for foundations. Short-bored, that is short length, piles are cast in holes augered by hand or machine. The piles support reinforced concrete ground beams off which walls are raised as illustrated in Fig. 13. The spacing of the piles depends on the loads and the plan of the building. Piles are cast below angles and intersections of external and internal loadbearing walls and spaced between these points where the loads and the span of ground beam requires additional piles.

Foundations on sloping sites

The foundation of walls on sloping sites may be at one level or stepped as illustrated in Fig. 14.

Where the slope is slight the foundation may be at one level with the floor raised above the highest ground level. Where there is a greater slope it is usual practice to cut and fill so that the wall at the highest point does not act as a retaining wall and there is no need to raise the ground floor above the highest point of the site as illustrated in Fig. 16. The term 'cut and fill' describes the operation of cutting into part of the higher part of the site and filling the remaining lower part with excavated material or with imported fill. It will be seen that the cutting is extended beyond the wall at the highest point to provide a drained dry area behind it.

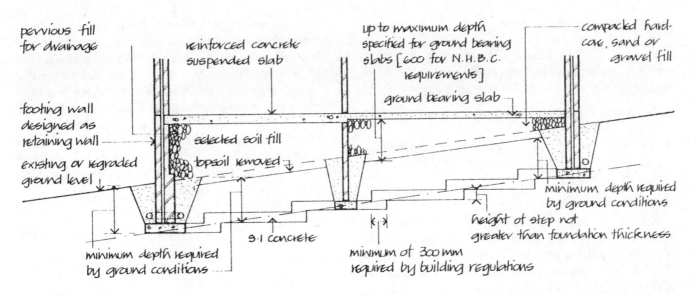

pervious fill for drainage

reinforced concrete suspended slab

footing wall designed as retaining wall

existing or regraded ground level

selected soil fill

topsoil removed

up to maximum depth specified for ground bearing slabs [600 for N.H.B.C. requirements]

ground bearing slab

compacted hard-core, sand or gravel fill

minimum depth required by ground conditions

height of step not greater than foundation thickness

minimum depth required by ground conditions

9.1 concrete

minimum of 300 mm required by building regulations

Fig. 15

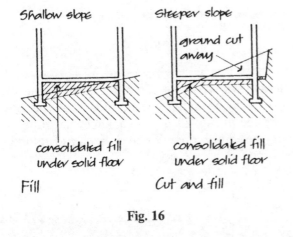

Shallow slope

Steeper slope

ground cut away

consolidated fill under solid floor

consolidated fill under solid floor

Fill

Cut and fill

Fig. 16

Where a building extends some distance up an appreciable slope it is usual to use stepped foundations to economise in excavation and foundation walling. The steps in the foundation should be uniform in height and not greater in height than the thickness of the foundation concrete and be a multiple of brick courses. The steps should extend over and unite with the lower foundation not less than twice the height of the step, by the thickness of the concrete foundation or 300 whichever is the greater, as illustrated in Fig. 15.

LAND DRAINAGE

Surface water (stormwater) is the term used for natural water, that is rainwater, from the surface of the ground including open ground such as fields,

paved areas and roofs. Rainwater that falls on paved areas and from roofs generally drains to surface water (stormwater) drains and thence to soakaways (see Volume 5), rivers, streams or the sea. Rainwater falling on natural open ground will in part lie on the surface of impermeable soils, evaporate to the air, run off to streams and rivers and soak into the ground. On permeable soils much of the rainwater will soak into the ground.

Ground water is that water held in soils at and below the water table (which is the depth at which there is free water below the surface). The level of the water table will vary seasonally being closest to the surface during rainy seasons and deeper during dry seasons when most evaporation to the air occurs.

In Part C of Schedule 1 to The Building Regulations 1985 is a requirement for subsoil drainage, to avoid passage of ground moisture to the inside of a building or to avoid damage to the fabric of the building.

In Approved Document C to the Regulations are provisions for the need for subsoil drainage where the water table can rise to within 0.25 m of the lowest floor and where the water table is high in dry weather and the site of the building is surrounded by higher ground.

Surface water (stormwater) drains

Paved areas are usually laid to falls to channels and gullies that drain to surface water drains (Volume 5).

Ground water (subsoil) drains

These are used to improve the run off of ground water to maintain the water table at some depth below the surface for the following reasons:

(a) to improve the stability of the ground
(b) to avoid surface flooding
(c) to alleviate or avoid dampness in basements
(d) to reduce humidity in the immediate vicinity of buildings.

Ground water, or land or field, drains are either open jointed or jointed, porous or perforated pipes of clayware, concrete, pitch fibres or plastic (see Volume 5). The pipes are laid in trenches to follow the fall of the ground, generally with branch drains discharging to a main drain in one of the following systems.

Natural system: The pipes are laid open jointed to follow the natural depressions and valleys of the land and connect to a main drain like tributaries to a river as illustrated in Fig. 17. This system is most used for draining open land for agriculture.

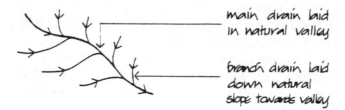

main drain laid in natural valley

branch drain laid down natural slope towards valley

Natural system of land drains

Fig. 17

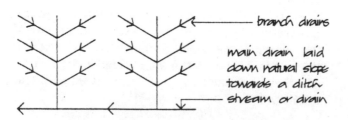

branch drains

main drain laid down natural slope towards a ditch stream or drain

Herring bone system of land drains

Fig. 18

Herringbone system: The pipes are laid in a series of herringbone patterns with branches, not exceeding 30 m, discharging at an angle to a main drain that follows the fall of the land. This, more regular system, may be used with jointed pipes for land drainage around buildings, as illustrated in Fig. 18.

Grid system: Branch and main drains are laid in a regular rectangular grid with the branches connecting to one side of main drains on the boundary of the land as illustrated in Fig. 19.

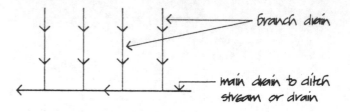

branch drain

main drain to ditch stream or drain

Grid system of land drains

Fig. 19

Fan shaped system: A fan of branches connect to a single main drain on the boundary of the site as illustrated in Fig. 20. This is a somewhat rough and ready system because of the difficulty of joining all the branches at one point.

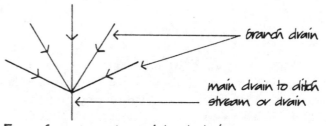

branch drain

main drain to ditch stream or drain

Fan shaped system of land drains

Fig. 20

Moat (cut-off) system: Drains laid on one or more sides of a building intercept the flow of ground water as illustrated in Fig. 21.

Ground water (land) drains are laid in trenches at depths of 0.6 m to 0.9 m in heavy soils and 0.9 m to 1.2 m in light soils. The nominal bore of the pipes is usually 75 and 100 for main drains and 65 or 75 for branches.

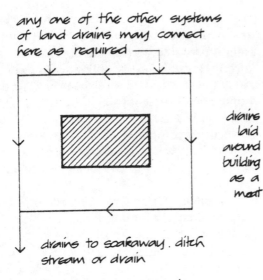

any one of the other systems of land drains may connect here as required

drains laid around building as a moat

drains to soakaway. ditch stream or drain

Moat or cut off system of land drains

Fig. 21

The drain trench is made wide enough to surround the pipes with clinker or is shaped, in cohesive soils, to accommodate the pipes as illustrated in Fig. 22. The pipes should be surrounded with clinker, gravel or rubble covered with inverted turf, brushwood or straw to keep fine soil from the pipes. The trench is backfilled with excavated spoil from the excavation. Where field drains are designed to collect both surface and ground water, the trench is filled with clinker, gravel or rubble to the surface as illustrated in Fig. 23.

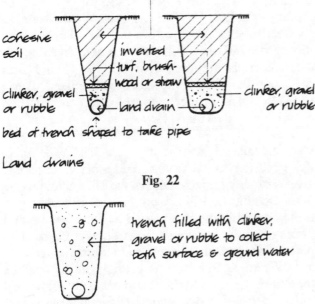

excavated material backfilled

cohesive soil

inverted turf, brush-wood or straw

clinker, gravel or rubble

clinker, gravel or rubble

land drain

bed of trench shaped to take pipe

Land drains

Fig. 22

trench filled with clinker, gravel or rubble to collect both surface & ground water

Fig. 23

For surface water drainage the French drain is often used. This is a trench filled with clinker, gravel or rubble as illustrated in Fig. 24.

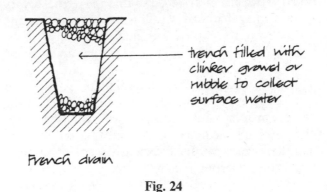

trench filled with clinker gravel or rubble to collect surface water

French drain

Fig. 24

CONCRETE

Concrete (see also Volume 4) is the name given to a mixture of particles of stone bound together with cement. Because the major part of concrete is of particles of broken stones and sand, it is termed the aggregate. The material which binds the aggregate is cement and this is described as the matrix.

Aggregate: The materials commonly used as the aggregate for concrete are sand and gravel. The grains of natural sand and particles of gravel are very hard and insoluble in water and can be economically dredged or dug from pits and rivers. The material dug from many pits and river beds consists of a mixture of sand and particles of gravel and is called 'ballast' or 'all-in aggregate'. The name ballast derives from the use of this material to load empty ships and barges. The term 'all-in aggregate' is used to describe the natural mixture of fine grains of sand and larger coarse particles of gravel.

All-in aggregate (ballast) is one of the cheapest materials that can be used for making concrete and was used for mass concrete work, such as large open foundations. The proportion of fine to coarse particles in an all-in aggregate cannot be varied and the proportion may vary from batch to batch so that it is not possible to control the mix and therefore the strength of concrete made with all-in aggregate. Accepted practice today is to make concrete for building from a separate mix of fine and coarse aggregate which is produced from ballast by washing, sieving and separating the fine from the coarse aggregate.

Fine aggregate is natural sand which has been washed and sieved to remove particles larger than 5 and coarse aggregate is gravel which has been crushed, washed and sieved so that the particles vary from 5 up to 50 in size. The fine and coarse aggregate are delivered separately. Because they have to be sieved, a prepared mixture of fine and coarse aggregate is more expensive than natural all-in aggregate. The reason for using a mixture of fine and coarse aggregate is that by combining them in the correct proportions, a concrete with very few voids or spaces in it can be made and this reduces the quantity of comparatively expensive cement required to produce a strong concrete.

Cement: The cement commonly used is ordinary Portland cement. It is manufactured by heating a mixture of finely powdered clay and limestone with water to a temperature of about 1300°C, at which the lime and clay fuse to form a clinker. This clinker is ground with the addition of a little gypsum to a fine powder of cement. Cement powder reacts with water and its composition gradually changes and the particles of cement bind together and adhere strongly to materials with which they are mixed. Cement hardens gradually after it is mixed with water. Some thirty minutes to an hour after mixing with water the cement is no longer plastic and it is said that the initial set has occurred. About ten hours after mixing with water, the cement has solidified and it increasingly hardens until some seven days after mixing with water when it is a dense solid mass.

Water-cement ratio: The materials used for making concrete are mixed with water for two reasons. Firstly to cause the reaction between cement and water which results in the cement acting as a binding agent and secondly to make the materials of concrete sufficiently plastic to be placed in position. The ratio of water to cement used in concrete affects its ultimate strength, and a certain water-cement ratio produces the best concrete. If too little water is used the concrete is so stiff that it cannot be compacted and if too much water is used the concrete does not develop full strength. The amount of water required to make concrete sufficiently plastic depends on the position in which the concrete is to be placed. The extreme examples of this are concrete for large foundations which can be mixed with comparatively little water and yet be consolidated, and concrete to be placed inside formwork for narrow reinforced concrete beams where the concrete has to be comparatively wet to be placed. In the first example, as little water is used, the proportion of cement to aggregate can be as low as say 1 part of cement to 9 of aggregate and in the second, as more water has to be used, the proportion of cement to aggregate has to be as high as say 1 part of cement to $4\frac{1}{2}$ of aggregate. As cement is expensive compared with aggregate it is usual to use as little water and therefore cement as the necessary plasticity of the concrete will allow.

Proportioning materials: The materials used for mass concrete for foundations were often measured out by volume, the amount of sand and coarse aggregate being measured in wooden boxes constructed for the purpose. This is a crude method of measuring the materials because it is laborious to have to fill boxes and then empty them into mixers and no account is taken of the amount of water in the aggregate. The amount of water in aggregate affects the finished concrete in two ways: (a) if the aggregate is very wet the mix of concrete may be too weak, have an incorrect ratio of water to cement and not develop full strength and, (b) damp sand occupies a greater volume than dry. This increase in volume of wet sand is termed bulking. The more accurate method of proportioning the materials for concrete is to measure them by weight. The materials used in reinforced concrete are commonly weighed and mixed in large concrete mixers. It is not economical for builders to employ expensive concrete mixing machinery for small buildings and the concrete for foundations, floors and lintels is usually delivered to site ready mixed, except for small batches that are mixed by hand or in a portable petrol driven mixer. The materials are measured out by volume and providing the concrete is thoroughly mixed, is not too wet and is properly consolidated the finished concrete is quite satisfactory.

Concrete mixes: British Standard 5328: 1981 Specifying concrete, including ready-mixed concrete, gives a range of mixes. One range of concrete mixes in the Standard, ordinary prescribed mixes, is suited to general building work such as foundations and floors. These prescribed mixes should be used in place of the traditional nominal volume mixes such as 1:3:6 cement, fine and coarse aggregate by volume, that have been used in the past. The prescribed mixes, specified by dry weight of aggregate, used with 100 kg of cement, provide a more accurate method of measuring the proportion of cement to aggregate and as they are measured against the dry weight of aggregate, allow for close control of the water content and therefore the strength of the concrete.

The prescribed mixes are designated by letters and numbers as, C7.5P, C10P, C15P, C20P, C25P and C30P. The letter C stands for 'Compressive', the letter P for 'Prescribed' and the number indicates the 28-day characteristic cube crushing strength in newtons per square millimetre (N/mm^2) which the concrete is expected to attain. The prescribed mix specifies the proportions of the mix to give an indication of the strength of the concrete sufficient for most building purposes, other than designed reinforced concrete work.

Table 5 equates the old nominal volumetric mixes of cement and aggregate with the prescribed mixes and indicates uses for these mixes.

Table 5. Concrete Mixes

Nominal volume mix	BS5328 mix	Uses
1:8 all-in } 1:3:6	C7.5P	Foundations Site concrete
1:6 all-in } 1:2½:5	C10P C15P	Foundations Site concrete
1:2:4	C20P	Min. quality plain struc. concrete and protected reinf. concrete
1:1½:3	C30P	Reinforced concrete

Ready-mixed concrete: The very many ready-mixed concrete plants in the United Kingdom are able to supply to all but the most isolated building sites. These plants prepare carefully controlled concrete mixes which are delivered to site by lorries on which the concrete is churned to delay setting. Because of the convenience and the close control of these mixes, much of the concrete used in building today is provided by ready-mixed suppliers. To order ready-mixed concrete it is only necessary to specify the prescribed mix, for example C10P, the cement, type and size of aggregate and the workability, that is medium or high workability, depending on the ease with which the concrete can be placed and compacted.

There are water soluble sulphates in some soils, such as plastic clay, which react with ordinary cement and in time will weaken concrete. It is usual practice, therefore, to use one of the sulphate-resistant cement for concrete in contact with sulphate bearing soils.

Portland blast-furnace cement: is more resistant to the destructive action of sulphates than ordinary Portland cement and is often used for concrete foundations in plastic clay subsoils. This cement is made by grinding a mixture of ordinary Portland cement with blast-furnace slag. Alternatively another type of cement know as 'sulphate resisting cement' is often used.

Sulphate resisting Portland cement: This cement has a reduced content of the aluminates that combine with soluble sulphates in some soils and is used for concrete in contact with those soils.

SITE PREPARATION

Turf and vegetable top soil is removed from the ground to be covered by a building, to a depth sufficient to prevent later growth. Tree and bush roots, that might encourage later growth, are grubbed up and any pockets of soft compressible material, that might affect the stability of the building, are removed. The reasons for removing this vegetable soil are firstly to prevent plants, shrubs or trees from attempting to grow under the concrete. In growing, even the smallest of plant life exerts considerable pressure – which would quite quickly rupture the concrete oversite. The second reason for removing the vegetable top soil is that it is generally soft and compressible and readily retains moisture which would cause concrete over it to be damp at all times. The depth of vegetable top soil varies and on some sites it may be necessary to remove 300 or more vegetable top soil.

In section 1 of Approved Document C to The Building Regulations 1985 is a list of possible contaminants in or on ground to be covered by a building, that may be a danger to health or safety. Building sites that may be likely to contain contaminants can be identified from planning records or local knowledge of previous uses. Sites previously used as asbestos, chemical or gas works, metal works, munitions factories, nuclear installations, oil stores, railway land, sewage works and land fill are some examples given.

Where there are signs of possible contaminants in or on ground to be built on, as detailed in Table 2 of section C, the Environmental Health Officer should be informed. Depending on the nature of the contaminants various actions should be taken such as sealing or removal of contaminant or contaminated ground to a depth of one metre and subsequent filling with inert material or remedial measures to reduce the risks of danger to health to acceptable levels, with the help of expert advice.

OVERSITE CONCRETE (concrete oversite)

Resistance to ground moisture

Up to about the middle of the nineteenth century the ground floor of most buildings was formed on compacted soil or dry fill on which was laid a surface of stone flagstones, brick or tile or a timber boarded floor nailed to battens bedded in the compacted soil or fill. In lowland areas and on poorly drained soils most of these floors were damp and cold underfoot.

A raised timber ground floor was used to provide a comparatively dry floor surface of boards, nailed to joists, raised above the packed soil or dry fill. To minimise the possibility of the joists being affected by rising damp it was usual to ventilate the space below the raised floor. The inflow of cold outside air for ventilation tended to make the floor cold underfoot.

When Portland cement was first continuously produced, towards the end of the nineteenth century, it became practical to cover the site of buildings with a layer of concrete as a solid level base for floors and as a barrier to rising damp. From the early part of the twentieth century it became accepted practice to cover the site of buildings with a layer of concrete some 100 thick, the concrete oversite or oversite concrete. At the time, many ground floors of houses were formed as raised timber floors on oversite concrete with the space below the floor ventilated against stagnant damp air.

With the shortage of timber that followed the Second World War, the raised timber ground floor was abandoned and the majority of ground floors were formed as solid, ground supported floors with the floor finish laid on the concrete oversite. At the time it was accepted practice to form a continuous horizontal damp-proof course, some 150 above ground level, in all walls with foundations in the ground.

With the removal of vegetable top soil the level of the soil inside the building would be from 100 to 300 below the level of the ground outside. If a layer of concrete were then laid oversite its finished level would be up to 200 below outside ground level and up to 350 below the horizontal d.p.c. in walls. There would then be considerable likelihood of moisture rising through the foundation walls, to make the inside walls below the d.p.c. damp as illustrated in Fig. 25.

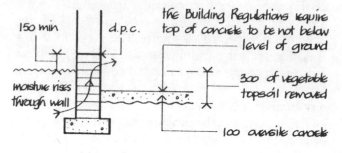

Diagram to illustrate need for hardcore

Fig. 25

It would, of course, be possible to make the concrete oversite up to 450 thick so that its top surface was level with the d.p.c. and so prevent damp rising into the building. But this would be unnecessarily expensive. Instead, a layer of what is known as hardcore is spread oversite, of sufficient thickness to raise the level of the top of the concrete oversite to that of the d.p.c. in walls. The purpose of the hardcore is primarily to raise the level of the concrete oversite for solid, ground supported floors.

The layer of concrete oversite will serve as a reasonably effective barrier to damp rising from the ground largely by absorbing moisture from below. The moisture retained in the concrete will tend to make solid floor finishes cold underfoot and may adversely affect timber floor finishes. During the second half of the twentieth century it became accepted practice to form a waterproof membrane under, in or over the oversite concrete as a barrier to rising damp, against the cold underfoot feel of solid floors and to protect floor finishes. Having accepted the use of a damp-proof membrane it was then logical to unite this barrier to damp to the damp-proof course in walls, by forming them at the same level or by running a vertical d.p.c. up from the lower membrane to unite with the d.p.c. in walls.

Even with the damp-proof membrane there is some appreciable transfer of heat from heated buildings through the concrete and hardcore to the cold ground below. In the 1989 revision of Approved

Document L to The Building Regulations 1985, is the inclusion of provision for insulation to ground floors for the conservation of fuel and power. The new requirement can be met by a layer of insulating material under the site concrete, under a floor screed or under boarded or sheet floor finishes to provide a maximum U value of 0.45 for the floor.

The requirements from Part C of Schedule 1 to The Building Regulations 1985 for the resistance of the passage of moisture to the inside of the building through floors are met if the ground is covered with dense concrete laid on a hardcore bed and a damp-proof membrane. The concrete should be at least 100 thick and composed of 50 kg of cement to not more than 0.11 m^3 of fine aggregate and 0.16 m^3 of coarse aggregate or BS 5326, mix C10P. The hardcore bed should be of broken brick or similar inert material, free from materials including water soluble sulphates in quantities which could damage the concrete. A damp-proof membrane, above or below the concrete, should be continuous with the damp-proof course in walls.

It is practice on building sites to first build external and internal load bearing walls from the concrete foundation up to the level of the damp-proof course, above ground, in walls. The hardcore bed and the oversite concrete are then spread and levelled within the external walls.

If the hardcore is spread over the area of the ground floor and into excavations for foundations and soft pockets of ground that have been removed and the hardcore is thoroughly consolidated by ramming, there should be very little consolidation settlement of the concrete ground supported floor slab inside walls. Where a floor slab has suffered settlement cracking, it has been due to an inadequate hardcore bed, poor filling of excavation for trenches or ground movement due to moisture changes. It has been suggested that the floor slab be cast into walls for edge support. This dubious practice, which requires edge formwork support of slabs at cavities, will have the effect of promoting cracking of the slab, that may be caused by any slight consolidation settlement. Where appreciable settlement is anticipated it is best to reinforce the slab and build it into walls as a suspended reinforced concrete slab.

Hardcore

This is the name given to the infill of materials such as broken bricks, stone or concrete, which are hard and do not readily absorb water or deteriorate. This hardcore is spread over the site within the external walls of the building to such thickness as required to raise the finished surface of the site concrete. The hardcore should be spread until it is roughly level and rammed until it forms a compact bed for the oversite concrete. This hardcore bed is usually from 100 to 300 thick.

The hardcore bed serves as a solid working base for building and as a bed for the concrete oversite. If the materials of the hardcore are hard and irregular in shape they will not be a ready path for moisture to rise by capillarity. Materials for hardcore should, therefore, be clean and free from old plaster or clay which in contact with broken brick or gravel would present a ready narrow capillary path for moisture to rise.

The materials used for hardcore should be chemically inert and not appreciably affected by water. Some materials used for hardcore, for example colliery spoil, contain soluble sulphate that in combination with water combine with cement and cause concrete to disintegrate. Other materials such as shale may expand and cause lifting and cracking of concrete. A method of testing materials for soluble sulphate is described in *Building Research Station (BRS) Digest 174*.

The materials used for hardcore are:

Brick or tile rubble: Clean, hard broken brick or tile is an excellent material for hardcore. Bricks should be free of plaster. On wet sites the bricks should not contain appreciable amounts of soluble sulphate.

Concrete rubble: Clean, broken, well-graded concrete is another excellent material for hardcore. The concrete should be free from plaster and other building materials.

Gravel and crushed hard rock: Clean, well-graded gravel or crushed hard rock are both excellent, but somewhat expensive materials for hardcore.

Chalk: Broken chalk is a good material for hardcore providing it is protected from expansion due to frost. Once the site concrete is laid it is unlikely to be affected by frost.

Pulverised fuel ash (PVA) or fly ash: The fine waste material from coalburning power stations is a good material for hardcore.

Blast-furnace slag: from the making of iron is a strong material which is used as hardcore.

Colliery spoil: is used as a material for hardcore. Unburnt colliery spoil is preferable to burnt spoil as the former has a lower soluble sulphate content.

Before the concrete is laid it is usual to blind the top surface of the hardcore. The purpose of this is to prevent the wet concrete running down between the lumps of broken brick or stone, as this would make it easier for water to seep up through the hardcore and would be wasteful of concrete. To blind, or seal, the top surface of the hardcore a thin layer of very dry coarse concrete can be spread over it, or a thin layer of coarse clinker or ash can be used. This blinding layer, or coat, will be about 50 thick, and on it the site concrete is spread and finished with a true level top surface. Figure 26 is an illustration of hardcore, blinding and concrete oversite. Even with a good hardcore bed below the site concrete a dense hard floor finish, such as tiles, may be slightly damp in winter and will be cold underfoot. To reduce the coldness experienced with some solid ground floor finishes it is good practice to form a continuous damp-proof membrane in the site concrete.

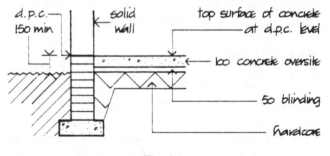

Fig. 26

DAMP-PROOF MEMBRANE

Resistance to ground moisture

Concrete is spread oversite as a barrier to moisture rising from the ground. Concrete is to some degree permeable to water and will absorb moisture from the ground, a damp oversite concrete slab will be cold and draw appreciable heat from rooms if it is to be maintained at an equable temperature.

A requirement from Part C of Schedule 1 to The Building Regulations 1985 is that floors shall adequately resist the passage of moisture to the inside of

the building. As concrete is permeable to moisture, it is generally necessary to use a damp-proof membrane under, in or on top of ground supported floor slabs as an effective barrier to moisture rising from the ground. The membrane should be continuous with the damp-proof course in walls, as a barrier to moisture rising between the edges of the concrete slab and walls.

A damp-proof membrane should be impermeable to water either in liquid or vapour form and be tough enough to withstand possible damage during the laying of screeds, concrete or floor finishes. The damp-proof membrane may be on top, sandwiched in or under the concrete slab.

Being impermeable to water the membrane will delay the drying out of wet concrete to ground if it is under the concrete, or screed to concrete if it is on top of the concrete.

Surface damp-proof membranes: Floor finishes such as pitch mastic and mastic asphalt that are impermeable to water act as a combined damp-proof membrane and floor finish as illustrated in Fig. 27. Adhesives of hot soft bitumen or coal-tar pitch in a continuous layer for wood block floor finishes may serve as an effective surface membrane.

Where timber fillets are set into the concrete slab as a fixing for boards, they should be treated with an effective preservative, unless they are above the damp-proof membrane.

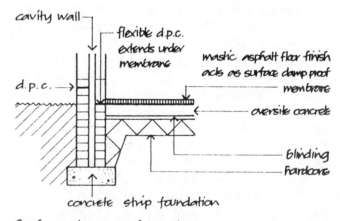

Surface damp-proof membrane

Fig. 27

Damp-proof membranes below a floor screed: Where neither the floor finish itself nor the adhesive for a floor finish act as a damp-proof membrane, a damp-proof membrane may be sandwiched between the concrete oversite and the screed as illustrated in Fig. 28. The membrane will delay the drying out of wet screed to the concrete below and will prevent adhesion of the screed to the surface below. The screed should be at least 65 thick to minimise the possibility of folding and cracking of the screed, due to unrestrained shrinkage.

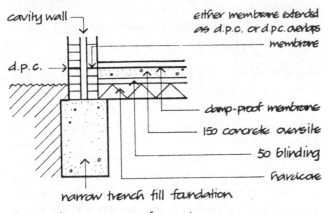

Sandwich damp-proof membrane

Fig. 28

Damp-proof membrane below site concrete: The membrane is laid or spread on a blinding of comparatively dry concrete, clinker or ash spread over the hardcore as illustrated in Fig. 29. These damp-proof membranes will delay the drying out of concrete to the extent of one month for each 25 thickness of concrete.

The advantage of a damp-proof membrane under the site concrete is that it is protected during subsequent building operations. The under slab membrane is used where there is under floor heating.

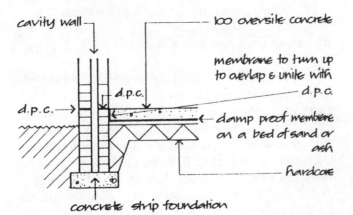

Below site concrete damp-proof membrane

Fig. 29

Materials for damp-proof membranes

Hot pitch or bitumen: A continuous layer of hot applied coal-tar pitch or soft bitumen is poured on the surface and spread to a thickness of not less than 3. In dry weather, concrete is ready for the membrane three days after placing concrete. The surface of the concrete should be brushed to remove dust and primed with a solution of coal-tar pitch or bitumen solution or emulsion. The pitch is heated to 35°C to 45°C and the bitumen to 50°C to 55°C.

Bitumen solution, bitumen/rubber emulsion or tar/ rubber emulsion: These cold applied solutions are brushed on to the surface of concrete in three coats to a finished thickness of not less than 2.5 mm, allowing each coat to harden before the next is applied.

Mastic asphalt or pitch mastic: These materials are spread hot and finished to a thickness of at least 12.5 mm. This expensive damp-proof membrane is used where there is appreciable water pressure under the floor.

Polythene and polyethylene sheet: Polythene and polyethylene sheet is used as a damp-proof membrane with oversite concrete for all but severe conditions of dampness. It is recommended that the sheet should be at least 0.25 mm thick (1000 gauge) to avoid damage during installation. Many suppliers recommend a thickness of 0.3 mm (1200 gauge). The sheet is supplied in rolls 4 m wide by 25 m long. When used under concrete oversite the sheet should be laid on a blinding layer of sand or compacted fuel ash spread over the hardcore.

The sheets are spread over the blinding and lapped 150 at joints and continued across surrounding walls, under the d.p.c. for the thickness of the wall.

Where site conditions are reasonably dry and clean, the overlap joints between the sheets are sealed with mastic or mastic tape between the overlapping sheets and the joint completed with a polythene jointing tape as illustrated in Fig. 30.

Where site conditions are too wet to use mastic and tape, the joint is made by welting the overlapping sheets with a double welted fold as illustrated in Fig. 30, and this fold is kept in place by weighing it down with bricks or securing it with tape until the screed or concrete has been placed.

Where the level of the membrane is below that of the d.p.c. in adjacent walls, the sheet can be dressed

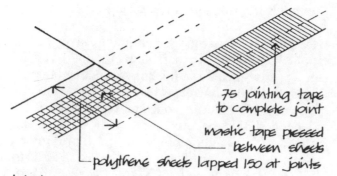

75 jointing tape to complete joint

mastic tape pressed between sheets

polythene sheets lapped 150 at joints

Jointing laps in polythene sheet

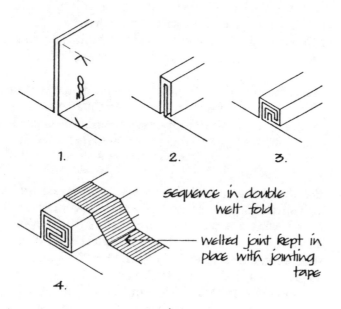

1. 2. 3.

sequence in double welt fold

welted joint kept in place with jointing tape

4.

Double welted fold joint in polythene sheet

Fig. 30

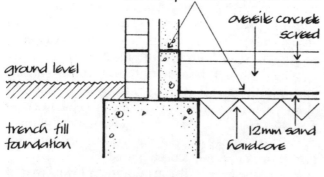

polythene damp proof membrane turned up & continued as D.P.C.

oversite concrete screed

ground level

trench fill foundation

12mm sand

hardcore

Fig. 31

up against the adjacent walls and then continued over the walls below the d.p.c., as illustrated in Fig. 31. This is a difficult operation, particularly at angles and it is advantageous to form the membrane at the same level as the d.p.c., in walls.

Provided it is protected against damage during laying, subsequent building operations and in the spreading or laying of screed or concrete, these sheets act as a very effective and economical barrier against rising damp.

Bitumen sheet: Sheets of bitumen with hessian, fibre or asbestos base are spread on the concrete oversite or on a blinding of stiff concrete below the concrete, in a single layer with the joints between adjacent sheets lapped 75. The joints are then sealed with a gas torch which melts the bitumen in the overlap of the sheets sufficient to bond them together. Alternatively the lap is made with hot bitumen spread between the overlap of the sheets which are then pressed together to make a damp-proof joint. The bonded sheets may be carried across adjacent walls as a d.p.c., or up against the walls and then across as d.p.c. where the membrane and d.p.c. are at different levels.

The polythene or polyester film and self-adhesive rubber/bitumen compound sheets, described in Volume 4 under 'Tanking', can also be used as damp-proof membranes, with the purpose cut, shaped cloaks and gussets for upstand edges and angles. This type of membrane is particularly useful where the membrane is below the level of the d.p.c., in walls.

Bitumen sheets are fairly readily damaged on building sites and should be covered for protection as soon as possible by the screed or site concrete.

Resistance to the passage of heat

The 1989 revision of the requirements of paragraph L in Schedule 1 to The Building Regulations 1985 and Approved Document L include provision for insulation to some ground floors. The requirement is that ground floors should have a maximum insulation value (U value) of 0.45 W/m^{2}K. Some ground floor slabs that are larger than 10 m in both length and breadth may not need the addition of an insulating layer to provide the U value of 0.45.

Of the heat that is transferred through a solid, ground supported floor a significant part of the transfer occurs around the perimeter of the floor to the ground below and foundation walls and ground around the edges of the floor, so that the cost of insulating the whole floor is seldom justified. Insulation around or under the edges of a solid floor will significantly reduce heat losses to the extent that overall insulation is unnecessary.

In the CIBS guide to the thermal properties of building structures, the U value of an uninsulated solid floor 20 × 20 m on plan, with four edges exposed, is given as 0.36 and one 10 × 10 m as 0.62. The 20 × 20 floor has a U value below that in the revised requirement of The Building Regulations and will not require insulation. The U value of the 10 m² floor can be reduced by the use of edge insulation. With edge insulation a metre deep all round or under the floor, the U value can be reduced to 0.48, which is somewhat higher than the U value in the revised requirement of The Building Regulations and may necessitate some small overall insulation. This is the basis for the assumption that floor slabs that are larger than 10 m in both length and width may not need an overall insulation layer.

To reduce heat losses through thermal bridges around the edges of solid floors that do not need overall insulation, and so minimise problems of condensation and mould growth, it may be wise to build in edge insulation particularly where the wall insulation is not carried down below the ground floor slab. Edge insulation is formed either as a vertical strip between the edge of the slab and the wall or under the slab around the edges of the floor as illustrated in Fig. 32. The depth or width of the strips of insulation vary from 0.25 m to 1 m and the thickness of the insulation will be similar to that needed for overall insulation.

The only practical way of improving the insulation of a solid ground floor to the required U value is to add a layer of some material with a high insulation value to the floor. The layer of insulation may be laid below a chipboard or plywood panel floor finish or below a timber boarded finish (see Chapter 3) or below the screed finish to a floor or under the concrete floor slab. With insulation under the screed or slab it is important that the density of the insulation board is sufficient to support the load of the floor itself and imposed loads on the floor. A density of at least 16 kg/m³ is recommended for domestic buildings.

The advantage of laying the insulation below the floor slab is that the high density slab, which warms and cools slowly (slow thermal response) in response

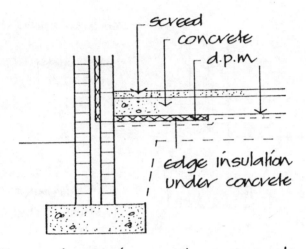

Edge insulation under concrete

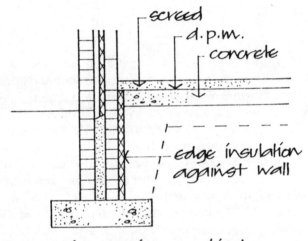

Edge insulation vertical

Perimeter insulation to ground supported slab.

Fig. 32

to changes in temperature of the constant low output heating systems, will not lose heat to the ground. The damp-proof membrane may be laid under or over the insulation layer or under the floor screed. The damp-proof membrane should be under insulation that absorbs water and may be over insulation with low water absorption and high resistance to ground contaminants.

With the insulation layer and the d.p.m. below the concrete floor slab it is necessary to continue the d.p.m. up vertically around the edges of the slab to unite with the d.p.c. in walls as illustrated in Fig. 33.

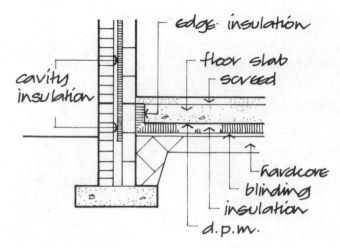

D.p.m. over insulation under floor slab.

Fig. 33

One method of determining the required thickness of insulation is to use a thickness of insulation related to the U value of the chosen insulation material, as for example, thicknesses of 25 for a U value of 0.02, 37 for 0.03, 49 for 0.04 and 61 for 0.05, ignoring the inherent resistance of the floor.

Another more exacting method is to calculate the required thickness related to the actual size of the floor and its uninsulated U value, taken from a table in the CIBS guide to the thermal properties of building structures. For example, from the CIBS table the U value of a solid floor 10×6 m, with four edges exposed is 0.74.

The inherent resistance of the floor	$= \dfrac{1}{0.74}$	$= 1.36 \text{ m}^2\text{K/W}$
Thermal resistance required	$= \dfrac{1}{0.45}$	$= 2.22 \text{ m}^2\text{K/W}$
Additional resistance required	$= 2.22 - 1.36$	
	$= 0.86 \text{ m}^2\text{K/W}$	
Assume an insulant with U value 0.04		
Thickness of insulation layer required	$= 0.86 \times 0.04 \times 1000$	
	$= 34.4 \text{ mm (49)}$	
or with an insulation value of 0.03	$= 25.8 \text{ mm (37)}$	

These thicknesses are appreciably less than those given by the first method, shown in brackets.

Where the wall insulation is in the cavity or on the inside face of the wall it is necessary to avoid a cold bridge across the foundation wall and the edges of

the slab, by fitting insulation around the edges of the slab or by continuing the insulation down inside the cavity as illustrated in Fig. 38. An advantage of fitting the d.p.m. above the insulation is that it can be used to secure the upstand edge insulation in place while concrete is being placed.

The disadvantage of the d.p.m. being below the concrete floor slab is that it will prevent the wet concrete drying out below and so lengthen the time required for it to adequately dry out, to up to six months. A concrete floor slab that has not been sufficiently dried out may adversely affect water sensitive floor finishes such as wood.

The advantage of laying the insulation layer under the screed is that it can be laid inside a sheltered building on a dried slab after the roof is finished and that the d.p.m., whether over or under the insulation layer, can more readily be joined to the d.p.c. in walls, as illustrated in Fig. 34. Where the wall insulation is in the cavity continue it down below the floor slab to minimise the cold bridge across the wall to the screed as illustrated in Fig. 34.

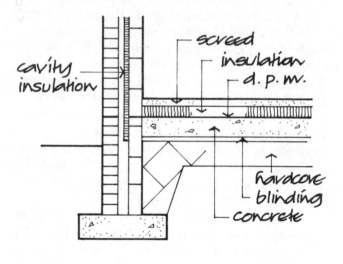

D.p.m. under insulation below screed

Fig. 34

If the d.p.m. is laid below the insulation it is necessary to spread a separating layer over the insulation to prevent wet screed running into the joints between the insulation boards. The separating layer should be building paper or 500 gauge polythene sheet.

To avoid damage to the insulation layer and the d.p.m. it is necessary to take care in tipping, spreading and compacting wet concrete or screed. Scaffold boards should be used for barrowing and tipping concrete and screed and a light mesh of chicken wire can be used over separating layers or d.p.m's over insulation under screeds as added protection.

Materials for underfloor insulation

Any material used as an insulation layer to a solid, ground supported floor must be sufficiently strong and rigid to support the weight of the floor or the weight of the screed and floor loads without undue compression and deformation. To meet this requirement one of the rigid board or slab insulants is used. The thickness of the insulation is determined by the nature of the material from which it is made and the construction of the floor, to provide the required U value.

Some insulants absorb moisture more readily than others and some insulants may be affected by ground contaminants. Where the insulation layer is below the concrete floor slab, with the d.p.m. above the insulation one of the insulants with low moisture absorption characteristics should be used.

The materials commonly used for floor insulation are rockwool slabs, extruded polystyrene, cellular glass and rigid polyurethane foam boards.

DAMP-PROOF COURSES

Resistance to ground moisture

The function of a damp-proof course is to act as a barrier to the passage of moisture or water between the parts separated by the damp-proof course. The movement of moisture or water may be upwards in the foundation of walls and ground floors, downwards in parapets and chimneys or horizontal where a cavity wall is closed at the jambs of openings.

One of the functional requirements of walls (see Chapter 2), is resistance to moisture. A requirement of Part C of Schedule 1 to The Building Regulations 1985, is that walls shall adequately resist the passage of moisture to the inside of the building. To meet this requirement it is necessary to form a barrier to moisture rising from the ground in walls. This barrier is the horizontal, above ground, damp-proof course.

Damp-proof courses above ground

There should be a continuous horizontal damp-proof course above ground in walls whose foundations are in contact with the ground, to prevent moisture from the ground rising through the foundation to the wall above ground, which otherwise would make wall surfaces damp and damage wall finishes. The damp-proof course above ground should be continuous for the whole length and thickness of the wall and be at least 150 above finished ground level.

It is convenient to group the materials used for damp-proof courses as flexible, semi-rigid and rigid. Flexible materials such as metal, bitumen and polythene sheet can accommodate moderate settlement movement in a building which may fracture the semi-rigid material mastic asphalt and probably fracture the rigid materials brick and slate.

Materials for damp-proof courses above ground

Flexible d.p.c.'s

Lead for use as a d.p.c. should weigh not less than 19.5 kg/m² (Code No. 4, –1.8 mm thick). Lead is an effective barrier to moisture and water. It is liable to corrosion in contact with freshly laid lime or cement mortar and should be protected by a coating of bitumen or bitumen paint applied to the mortar surface and both surfaces of the lead. Lead is durable and flexible and can suffer distortion due to moderate settlement in walls without damage. Lead is an expensive material and is little used today other than for ashlar stonework or as a shaped d.p.c. in chimneys. Lead should be laid in rolls the full thickness of the wall or leaf of cavity walls and be lapped at joints along the length of the wall and at intersections at least 100 or the width of the d.p.c.

Copper as a d.p.c. should be annealed at least 0.25 mm thick and have a nominal weight of 2.28 kg/m². Copper is an effective barrier to moisture and water, it is flexible, has high tensile strength and can suffer distortion due to moderate settlement in a wall without damage. It is an expensive material and is little used today as a d.p.c. above ground. When used as a d.p.c., it may cause staining of wall surfaces due to the oxide that forms. It is spread on an even bed of mortar and lapped at least 100 or the width of the d.p.c. at running joints and intersections.

Bitumen d.p.c.: There are six types of bitumen d.p.c., as follows:

(a) bitumen damp-proof course with hessian base
(b) bitumen damp-proof course with fibre base
(c) bitumen damp-proof course with asbestos base
(d) bitumen damp-proof course with hessian base and lead
(e) bitumen damp-proof course with fibre base and lead
(f) bitumen damp-proof course with asbestos base and lead

Bitumen d.p.c.'s are reasonably flexible and can withstand distortion due to moderate settlement in walls without damage. They may extrude under heavy loads without affecting their efficiency as a barrier to moisture. Bitumen d.p.c.'s, which are made in rolls to suit the thickness of walls, are bedded on a level bed of mortar and lapped at least 100 or the width of the d.p.c. at running joints and intersections.

Bitumen is the material most used for d.p.c.'s today because it is at once economical, flexible, reasonably durable and convenient to lay. There is little to choose between the three bases, hessian, fibre or asbestos as a base for a bitumen d.p.c. above ground. The fibre base is cheaper but less tough than hessian and the asbestos base is more resistant to being squeezed out under heavy loads.

The lead cored d.p.c., with a lead strip weighing not less than 1.20 kg/m^2, joined with soldered joints, is more expensive than the bitumen alone types. It is generally used as the horizontal d.p.c. for houses.

The combination of a mortar bed, bitumen d.p.c., and the mortar bed over the d.p.c. for brickwork, makes a comparatively deep mortar joint that may look unsightly.

Polythene sheet: for use as a d.p.c. should be black low density polythene sheet of single thickness not less than 0.46 mm weighing approximately 0.48 kg/ m^2. Polythene sheet is flexible, can withstand distortion due to moderate settlement in a wall without damage and is an effective barrier against moisture. It is laid on an even bed of mortar and lapped at least the width of the d.p.c. at running joints and intersections. Being a thin sheet material, polythene makes a thinner mortar joint than a bitumen d.p.c., and is sometimes preferred for that reason.

Its disadvantage as a d.p.c., is that it is fairly

readily damaged by sharp particles in mortar or the coarse edges of brick.

Polymer-based sheets are thinner than bitumen sheets and are used where the thicker bitumen d.p.c. mortar joint would be unsightly. This d.p.c. material, which has its laps sealed with adhesive, may be punctured by sharp particles and edges.

Semi-rigid d.p.c.'s

Mastic asphalt spread hot in one coat to a thickness of 13 is a semi-rigid d.p.c., impervious to moisture and water. Moderate settlement in a wall may well cause a crack in the asphalt through which moisture or water may penetrate. It is an expensive form of d.p.c., which shows on the face of walls as a thick joint and it is very rarely used as a d.p.c.

Rigid d.p.c.'s

Slates: Two courses of slates laid breaking joint in cement mortar were used as a d.p.c. before the introduction of bitumen as a d.p.c. A slate d.p.c. is little used today because of its comparatively high cost, the thick joint formed by the slates and mortar and the rigidity of the material that cannot accommodate moderate settlement in a wall without fracture.

Two courses of slates at least 230 long were laid, breaking joint in cement mortar as illustrated in Fig. 35.

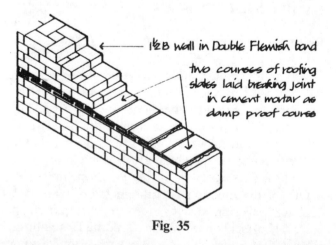

1½B wall in Double Flemish bond

two courses of roofing slates laid breaking joint in cement mortar as damp proof course

Fig. 35

Bricks: Two or three courses of dense engineering bricks laid in cement mortar serve as an effective barrier to moisture. This type of rigid d.p.c. is little used today because it is comparatively costly.

Damp-proof courses above ground should be at least 150 above the highest point of finished ground level to avoid the possibility of a build-up of material against the wall acting as a bridge for moisture from the ground as illustrated in Fig. 36.

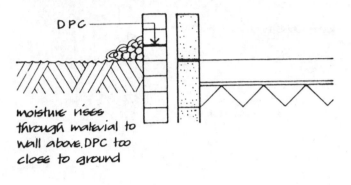

Fig. 36

All damp-proof courses above ground must extend the full width of the wall or the leaves of cavity walls and the edges of the d.p.c.'s must not be covered with mortar, pointing or rendering.

A damp-proof course in external walls should unite with the damp-proof membrane in, on or under the oversite concrete. This may be effected either by laying the membrane in the concrete at the same level as the d.p.c. in the wall or by uniting membrane and d.p.c., laid at different levels with a vertical d.p.c., as illustrated in Fig. 29. The vertical d.p.c. or membrane involves additional site labour and is difficult to make watertight at angles and intersections. In a cavity wall the d.p.c. in the inner leaf may be at a different level to that in the outer leaf so that it is level with the membrane in the oversite concrete as illustrated in Fig. 37.

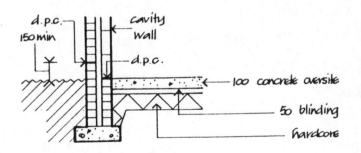

Fig. 37

The practical guidance given in Approved Document C to meeting the requirements in The Building Regulations 1985 for damp-proof courses, is that the cavity in a cavity wall should be taken down at least 150 below the level of the lowest damp-proof course or where the cavity stops at d.p.c. level, a damp-proof tray should be provided as illustrated in Fig. 38, to prevent rain or snow passing to the inner leaf.

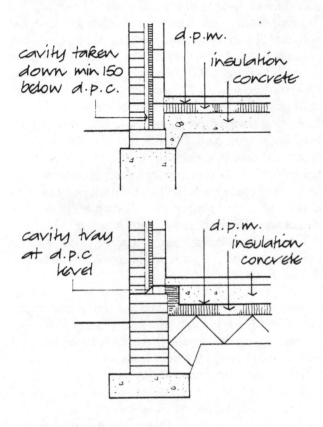

Protecting inner leaf of cavity wall

Fig. 38

EXCAVATION

The trenches which have to be dug for the foundations of walls are usually excavated by hand for single small buildings but where several houses are being built at the same time it is often economical to use mechanical trench diggers.

If the trenches are of any depth it may be necessary to erect temporary timber supports to stop the sides of the trench from falling in. The nature of the soils being excavated mainly determines the depth of trench for which timber supports to the sides should be used. Soft granular soils readily crumble and the sides of trenches in such soil may have to be supported for the full depth of the trench. The sides of trenches in clay soil do not usually require support for some depth, say up to 1.5, particularly in dry weather. In rainy weather if the bottom of the trench in clay soil gets filled with water, the water may wash out the clay from the sides at the bottom of the trench and then the whole of the sides above cave in. The purpose of temporary timbering supports to trenches is to uphold the sides of the excavation as necessary so as to avoid collapse of the sides which may endanger the lives of those working in the trench and to avoid the wasteful labour of constantly clearing falling earth from the trench bottoms. Whatever timbering is used there should be as few struts, that is horizontal members, fixed across the width of the trench as possible as these obstruct ease of working in the trench. All struts must be firmly secured so that they are not easily knocked out of position. The sides of deep trenches in compact soils such as clay should if necessary be timbered as shown in Fig. 39.

If the soil is soft, such as loam, more closely spaced timbering of the sides will be needed as shown in Fig. 40.

Dry granular soils such as sand and made-up ground may need closely spaced timbering to the sides (see Fig. 41).

The sizes of timbers shown in the drawings are for guidance only, as it is impossible to set out exact rules for determing the size of timbers required.

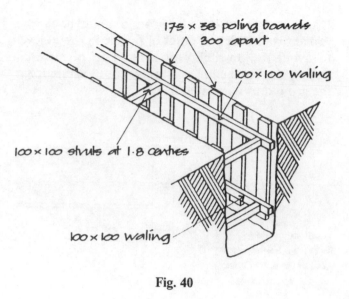

Fig. 40

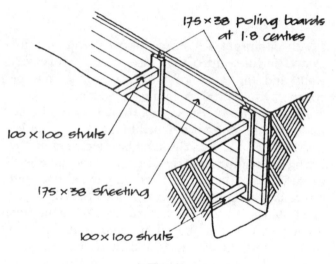

Fig. 41

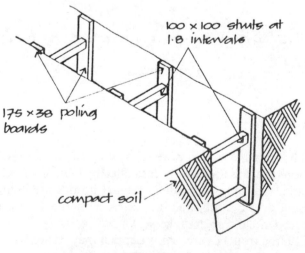

Fig. 39

28

CHAPTER TWO
WALLS

A wall is a continuous, usually vertical structure of brick, stone, concrete, timber or metal, thin in proportion to its length and height, which encloses and protects a building or serves to divide buildings into compartments or rooms.

Walls are defined as external or internal to differentiate functional requirements, and as load bearing or non-load bearing to differentiate structural requirements.

Load bearing walls are those that carry the loads of floors and roofs, in addition to their own weight, and non-load bearing, those that carry only their own weight. The word partition may be used to describe any continuous vertical structure that divides rooms or compartments. In practice, the word partition is generally used to describe a non-load bearing internal dividing wall.

Walls are of two types, solid or framed. A solid wall (sometimes called a masonry wall) is constructed either of brick, burned clay or blocks of stone or concrete laid in mortar with the blocks laid to overlap in some form of what is called bonding or as a monolith, that is, one solid uninterrupted material such as concrete which is poured wet and hardens into a solid monolith (one piece of stone). A solid wall of bricks or blocks may be termed a block (or masonry) wall, and a continuous solid wall of concrete, a monolithic wall. A frame wall is constructed from a frame of small sections of timber, concrete or metal joined together to provide strength and rigidity, over both faces of which, or between the members of the frame are fixed thin panels of some material to fulfill the functional requirements of the particular wall. Figure 42 is an illustration of the two types of wall.

Each of the two types of wall may serve as internal or external wall and as a load bearing or non-load bearing wall. Each of the two types of wall has different characteristics in fulfilling the functional requirements of a wall so that one type may have good resistance to fire but be a poor insulator against transfer of heat, and the other poor resistance to rain penetration yet good insulation against transfer of heat. There is no one material or type of wall that will

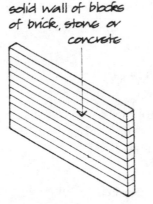
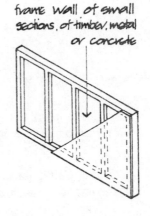

solid wall of blocks of brick, stone or concrete

frame wall of small sections of timber, metal or concrete

Solid wall

Frame wall

Fig. 42

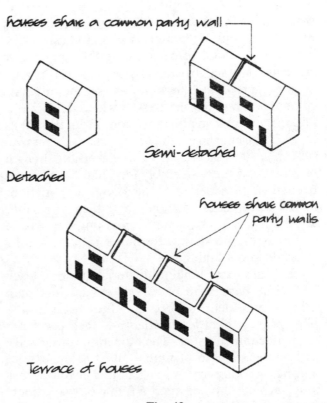

houses share a common party wall

Detached

Semi-detached

houses share common party walls

Terrace of houses

Fig. 43

fulfill all the functional requirements of a wall with maximum efficiency.

A traditional small building is built as a box of solid walls with openings for windows and doors in the form of detached, semi-detached or a terrace of houses, for example, as illustrated in Fig. 43. The solid block walls served as protection against wind and rain, to support floors and roofs and to some extent to contain heat within the building.

FUNCTIONAL REQUIREMENTS

The function of a wall is to enclose and protect a building or divide space within a building. The functional requirements of a wall are:

Strength and stability
Resistance to weather and ground moisture
Durability and freedom from maintenance
Fire resistance
Resistance to the passage of heat
Resistance to airborne and impact sound

Strength and stability

Strength: The strength of the materials used in construction is determined by the strength of a material in resisting compressive and tensile stress. The usual method of determining the compressive and tensile strength of a material is to subject samples of the material to tests to assess the ultimate compressive and tensile stress at which the material fails in compression and in tension. From these tests the safe working strengths of materials in compression and in tension are set. The safe working strength of a material is considerably less than the ultimate strength, to provide a safety factor against variations in the strength of materials and their behaviour under stress. The characteristic working strengths of materials to an extent, determines their use in the construction of buildings.

The traditional building materials timber, brick and stone have been in use since man first built permanent settlements, because of the ready availability of these natural materials and their particular strength characteristics. The moderate compressive and tensile strength of timber allied to the ease of cutting, working and joining timber members has long been used to construct a frame of walls, floors and roofs for houses.

The compressive strength of well-burned brick combined with the durability, fire resistance and appearance of the material commends it as a walling material for the more permanent buildings.

The sense of solidity and permanence and compressive strength of sound building stone made it the traditional walling material for many larger buildings.

Steel and concrete, which have been used in building since the Industrial Revolution, are used principally for their very considerable strength as the structural frame members of large buildings where the compressive and tensile strength of steel and the compressive strength of concrete, separately or in combination with steel, is used for both columns and beams.

In the majority of small buildings, such as houses, the compressive strength of brick and stone is rarely fully utilised because the functional requirements of stability and exclusion of weather dictate a thickness of wall in excess of that required for strength alone. To support the very modest loads on the walls of small buildings the thinnest brick or stone wall would be quite adequate.

Stability: The stability of a wall may be affected by foundation movement (see Chapter 1), eccentric loads, lateral forces (wind) and expansion due to temperature and moisture changes. Eccentric loads, that is those not acting on the centre of the thickness of the wall, such as from floors and roofs and lateral forces, such as wind, tend to deform and overturn walls. The greater the eccentricity of the loads and the greater the lateral forces, the greater the tendency of a wall to deform, bow out of the vertical and lose stability. To prevent loss of stability, due to deformation under loads, building regulations and structural design calculations set limits to the height to thickness ratios (slenderness ratios) to provide reasonable stiffness against loss of stability due to deformation under load.

To provide stiffness against deformation under load, lateral, that is horizontal restraint is provided by walls and roofs tied to the wall for stiffening up the height of the wall and by intersecting walls and piers that are bonded or tied to the wall as stiffening against deformation along the length of walls.

Irregular profile walls have greater stiffness against deformation than straight walls because of the buttressing effect of the angle of the zigzag, chevron, offset or serpentine profile of the walls illustrated in Fig. 44. The more pronounced the chevron, zigzag, offset or serpentine the wall the stiffer it will be.

Similarly the diaphragm and fin walls, described in Volume 3, are stiffened against overturning and loss of stability by the cross ribs or diaphragms built across the wide cavity of diaphragm walls and the fins or piers that are built and bonded to straight walls in fin wall construction.

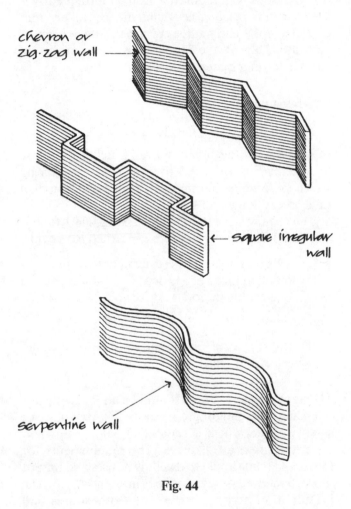

Fig. 44

Resistance to weather and ground moisture

The requirements of Part C of Schedule 1 of The Building Regulations 1985 are that walls should adequately resist the passage of moisture to the inside of the building. Moisture includes water vapour and liquid water. Moisture may penetrate a wall by absorption of water from the ground that is in contact with foundation walls or through rain falling on the wall.

To prevent water, which is absorbed from the ground by foundation walls, rising in a wall to a level where it might affect the inside of a building it is necessary to form a continuous, horizontal layer of some impermeable material in the wall. This impermeable layer, the damp-proof course, is built in, some 150 above ground level, to all foundation walls in contact with the ground and is joined to the damp-proof membrane in solid ground floors as described and illustrated in Chapter 1.

The ability of a wall to resist the passage of water to its inside face depends on its exposure to wind driven rain and the construction of the wall. The exposure of a wall is determined by its location and the extent to which it is protected by surrounding higher ground or sheltered by surrounding buildings or trees, from rain driven by the prevailing winds. In Great Britain the prevailing, warm westerly winds from the Atlantic Ocean cause more severe exposure to driving rain along the west coast of the country than do the colder easterly winds on the east coast.

British Standard 5628: Part 3 defines six categories of exposure as: very severe; severe; moderate/severe; sheltered/moderate; sheltered; and very sheltered. A map of Great Britain, published by the Building Research Establishment, shows contours of the variations of exposure across the country. The contour lines, indicating the areas of the categories of exposure, are determined from an analysis of the most severe likely spells of wind driven rain, occurring on average every three years, plotted on a 10 km grid. The analysis is based on the 'worst case' for each geographical area, where a wall faces open country and the prevailing wind, such as a gable end wall on the edge of a suburban site facing the prevailing wind or a wall of a tall building on an urban site rising above the surrounding buildings and facing the prevailing wind.

Where a wall is sheltered from the prevailing winds by adjacent high ground or surrounding buildings or trees the exposure can be reduced by one category in sheltered areas of the country and two in very severe exposure areas of the country. The small-scale and large-scale maps showing categories of exposure to driving rain provide an overall picture of the likely severity of exposure over the country. To estimate the likely severity of exposure to driving rain of the walls of a building on a particular site it is wise to take account of the categories of exposure shown on the maps, make due allowance for the overlap of categories around contour lines and obtain local knowledge of conditions from adjacent buildings and make allowance for shelter from high ground, trees and surrounding buildings.

The behaviour of a wall in excluding wind and rain will depend on the nature of the materials used in the construction of the wall and how they are put together. A wall of facing bricks laid in mortar will absorb an appreciable amount of the rain driven on to it so that the wall must be designed so that the rain is not absorbed to the inside face of the wall. This may be effected by making the wall of sufficient thickness, by applying an external facing of say rendering or slate hanging, or by building the wall as a cavity wall of two skins or leaves with a separating cavity.

A curtain wall of glass (see Volume 4) on the other hand will not absorb water through the impermeable sheets of glass so that driving rain will pour down the face of the glass and penetrate the joints between the sheets of glass and the supporting frame of metal or wood so that close attention has to be made to the design of these joints that at once have to be sufficiently resilient to accommodate thermal movement and at the same time compact enough to exclude wind and rain.

It is generally accepted practice today to construct walls of brick, stone or blocks as a cavity wall with an outer and an inner leaf or skin separated by a cavity of at least 50. The outer leaf will either be sufficiently thick to exclude rain or be protected by an outer skin of rendering or cladding of slate or tile and the inner leaf will be constructed of brick or block to support the weight of floors and roofs with either the inner leaf providing insulation against transfer of heat or the cavity filled with some thermal insulating material.

Durability and freedom from maintenance

The durability of a wall is indicated by the frequency and extent of the work necessary to maintain minimum functional requirements and an acceptable appearance. Where there are agreed minimum functional requirements such as exclusion of rain and thermal properties, the durability of walls may be compared through the cost of maintenance over a number of years. Standards of acceptable appearance may vary widely from person to person particularly with unfamiliar wall surface materials such as glass and plastic coated sheeting, so that it is difficult to establish even broadly-based comparative stand-ards of acceptable appearance. With the traditional wall materials there is a generally accepted view that a wall built of sound, well-burned bricks or wisely chosen stone 'looks good' so that there is to a considerable extent a consensus of acceptable appearance for the traditional walling materials.

In terms of the cost of maintenance over a number of years, the durability of a wall material can be assessed and included in the comparative cost of using various walling materials. Too infrequently is the cost of reasonable maintenance taken into account in the early stages of a building project or considered as part of the relative advantages of various walling materials and forms.

Fire-resistance

Fires in buildings generally start from a small source of ignition, the 'outbreak of fire', which leads to the 'spread of fire' followed by a steady state during which all combustible material burns steadily up to a final 'decay stage'.

The requirements of Part B of Schedule 1 to The Building Regulations 1985 are concerned to provide:

(a) adequate means of escape in case of fire
(b) maintenance of structural stability for a reasonable period

and to limit

(c) the spread of fire within and between buildings for a reasonable period

Means of escape: There should be adequate means of escape from buildings, in case of fire, to a place of safety outside, which is capable of being safely and effectively used at all times. The requirements for means of escape apply to dwelling houses and flats of more than three storeys and to offices and shops. The presumption here is that escape from houses and flats of up to three storeys, in a reasonable time, will be practical without the provision of special escape routes.

Means of escape from houses and flats of three or more storeys has to be provided by an adequate number of exits and protected escape routes from all floors three or more storeys above ground and from all shops and offices.

Structural stability in case of fire: The premature failure of the structural stability of a building is restricted by specifying a minimum period of fire resistance for the elements of the structure. The fire resistance of elements of the structure include requirements for:

(a) Stability – resistance to collapse of load bearing elements such as walls, beams, columns and floors.

(b) Integrity – resistance to fire penetration of fire separating elements such as casings to columns and beams.

(c) Insulation – resistance to the transfer of excessive heat through fire separating elements such as casings and linings to structural elements.

The minimum period of fire resistance of elements of structure depends on the use, height and size of the building or compartment. The required minimum period of resistance for basements is generally greater than that for ground and upper storeys because of the greater difficulty of dealing with fires in basements.

Spread of fire: Fire may spread within a building over the surface of materials covering walls and ceilings, through spaces between elements of the building and along spaces and voids in the structure such as spaces above suspended ceilings, voids behind linings and casings to ducts for pipes.

Fire may spread between buildings depending on the intensity of the fire, the distance between the buildings, the extent of openings in external walls and the protection given by the sides of the facing building.

Internal fire spread
(a) Surfaces: The spread of fire over internal surfaces, such as walls and ceilings, can be limited by the use of materials that have low rates of surface spread of flame.

(b) Compartments: The spread of fire can be limited by dividing a building into compartments of restricted floor area and cubic capacity, by means of compartment walls or compartment floors. The danger to life in case of fire in single storey buildings is less than that in multistorey buildings in which it is generally provident to form compartments to limit internal fire spread and reduce life risk associated with fire. Separating walls between adjoining buildings may serve as compartment walls to limit fire spread.

(c) Concealed spaces and fire stopping: Void spaces hidden above false ceilings, behind linings and in ducts for pipes provide a ready route for the spread of smoke and flame in buildings. The spread of smoke and fire in void spaces is hidden from view and presents a greater danger than a more obvious fire inside a building. The edges of void spaces are sealed to limit the spread of fire which might otherwise find its way around barriers to fire and lengthy voids such as pipe ducts, that run through buildings, are subdivided with fire stops at floor levels and around fire barriers.

External fire spread
The spread of fire between buildings is limited by control of the size of openings in external walls related to the space between a building and the boundary of the site facing the wall. By reference to the boundary of the site, the control will limit the external spread of fire to buildings that exist or may be built on adjacent sites. The term 'unprotected area' is used to embrace those parts of external walls that may contribute to the spread of fire between buildings and includes doors, windows and those parts of external walls that have less than the approved fire resistance for the wall and also external wall surfaces that may contribute to spread of flame.

The roofs of buildings may encourage spread of fire between buildings depending on the materials of the roof covering and the proximity of adjacent buildings.

Purpose groups: In Approved Document B to The Building Regulations 1985 is a designation of purpose groups such as dwelling house, flat, institutional, office, shop and industrial with a table of those types of building included in each group. The groups are listed in ascending order of risk of damage and danger from fire, related to the use of the building and its usual size. Dwelling houses are commonly comparatively small buildings in which the danger from fire is less than that in the usually larger office building. The provisions for means of escape, structural stability and limitation of spread of fire are set out separately for each purpose group together with limitations on size in each group.

Periods of fire resistance: Minimum periods of fire resistance are given in each purpose group for the elements of structure from 30 minutes for dwelling houses up to 4 hours for some offices. These minimum periods of fire resistance in hours, represent a notional period of time required after the outbreak of fire, for the occupants to escape to safety, which is related to the size and use of a building.

Resistance to the passage of heat

The traditional method of heating buildings was by burning wood or coal in open fireplaces in England and in freestanding stoves in much of northern Europe. The ready availability of wood and coal was adequate to the then modest demands for heating of the comparatively small population of those times. The highly inefficient open fire had the advantage of being a cosy focus for social life and the disadvantage of generating draughts of cold air necessary for combustion. The more efficient freestanding stove, which lacked the obvious cheery blaze of the open fire, was more suited to burning wood, the fuel most readily available in many parts of Europe.

The considerable increase in population that followed the Industrial Revolution and the accelerating move from country to town and city, increased the demand for the dwindling supplies of wood for burning. Coal became the principal fuel for open fires and freestanding stoves.

During the eighteenth century town gas became the principal source for lighting and by the nineteenth century had largely replaced solid fuels as the heat source for cooking. From about the middle of the twentieth century oil was used as the heat source for heating. Following the steep increase in the price of oil, town gas and later on natural gas was adopted as the fuel most used for heating.

Before the advent of oil and then gas as fuels for heating it was possible to heat individual rooms by means of solid fuel burning open fires or stoves and people accepted the need for comparatively thick clothing for warmth indoors in winter.

With the adoption of oil and gas as fuels for heating it was possible to dispense with the considerable labour of keeping open fires and stoves alight and the considerable area required to store an adequate supply of solid fuels. With the adoption of oil and gas as fuel for heating it was practical to heat whole buildings and there was no longer the inconvenience of cold corridors, toilets and bathrooms and the draughts of cold air associated with open fireplaces. The population increasingly worked in heated buildings, many in sedentary occupations, so that tolerance to cold diminished and the expectation of thermal comfort increased.

Of recent years the expectation of improved thermal comfort in buildings, the need to conserve natural resources and the increasing cost of fuels has led to the necessity for improved insulation against transfer of heat. To maintain reasonable and economical conditions of thermal comfort in buildings, walls should provide adequate insulation against excessive loss or gain of heat, have adequate thermal storage capacity and the internal face of walls should be at a reasonable temperature.

For insulation against loss of heat, lightweight materials with low conductivity are more effective than dense materials with high conductivity whereas dense materials have better thermal storage capcity than lightweight materials.

Where a building is continuously heated it is of advantage to use the thermal storage capacity of a dense material on the inside face of the wall with the insulating properties of a lightweight material behind it. Here the combination of a brick or dense block inner leaf, a cavity filled with some lightweight insulating material and an outer leaf of brick against penetration of rain is of advantage. Where buildings are intermittently heated it is important that inside faces of walls warm rapidly, otherwise if the inside face were to remain cold, the radiation of heat from the body to the cold wall face would make people feel cold. The rate of heating of smooth wall surfaces is improved by the use of low density, lightweight materials on or immediately behind the inside face of walls.

Where the material used in a wall is required to act as an insulator against transfer of heat the concern is in its capacity not as a conductor but as a resistor to transfer of heat and it is often convenient, therefore, to express its insulating property as resistivity which is the inverse of conductivity.

The loss or gain of heat through an element such as a wall is determined by its thermal transmittance or conductance which is expressed as its U value. Thermal transmittance depends on the temperature difference between the air on each side of the wall, the thermal conductance values of the materials of the wall and windows and on their relative areas. The U value is obtained from the sum of the thermal resistances of the component parts of the wall and by taking their reciprocal. Thus the thermal transmittance, or U value, is a measure of the conductance to transfer of heat, so that the higher the U value, the lower the insulation, and the lower the U value, the higher the insulation against transfer of heat.

Ventilation

The sensation of comfort is highly subjective and depends on the age, activity and to a large extent on the expectation of the subject. The young 'feel' cold less than the old and someone engaged in heavy manual work has less need of heating than another engaged in sedentary work. It is possible to provide conditions of thermal comfort that suit the general expectations of those living or working in a building. None the less some may 'feel' cold and others 'feel' hot.

For comfort and good health in buildings it is necessary to provide means of ventilation through air changes through windows or ventilators, that can be controlled, depending on wind speed and direction and outside air temperature, to avoid the sensation of 'stuffiness' or cold associated with too infrequent or too frequent air changes respectively. As with heating, the sensation of stuffiness is highly subjective. For general guidance a number of air changes per hour is recommended, depending on the activity common to rooms or spaces. One air change each hour for dwellings and three for kitchens and sanitary accommodation is recommended. The more frequent air changes for kitchen and sanitary accommodation is recommended to minimise condensation of moisture-laden, warm air on cold internal surfaces in those rooms.

Condensation is the effect of moisture from air collecting on a surface colder than the air, for example in a bathroom or kitchen where water from warm moisture-laden air condenses on to the cold surfaces of walls and glass. To minimise condensation, ventilation of the room to exchange moisture-laden air with drier outside air and good insulation of the inner face of the wall are required.

A consequence of the need for internal air change in buildings is that the heat source must be capable of warming the incoming air to maintain conditions of thermal comfort and the more frequent the air change the greater the heat input needed.

The major source of heat loss through walls is by window glass which is highly conductive to heat transfer. This heat loss can be reduced to some small extent by the use of double glazing. Most of the suppliers of double glazed windows provide one of the very effective air seals around all of the opening parts of their windows. These air seals are very effective in excluding the draughts of cold air that otherwise would penetrate the necessary gaps around opening windows and so serve to a large extent to reduce the heat loss associated with opening windows to an extent that they may reduce air changes to an uncomfortable level. There is a fine balance between the need for air change and the expectations of thermal comfort that receives too little consideration in the design of windows.

Resistance to airborne and impact sound

Sound is transmitted as airborne sound and impact sound. Airborne sound is generated as cyclical disturbances of air from, for example, a radio, that radiates from the source of the sound with diminishing intensity. The vibrations in the air caused by the sound source will set up vibrations in enclosing walls and floors which will cause vibrations of air on the opposite side of walls and floors.

Impact sound is caused by contact with a surface, as for example the slamming of a door or footsteps on a floor which sets up vibrations in walls and floors that in turn cause vibrations of air around them that are heard as sound.

The most effective insulation against airborne sound is a dense barrier such as a solid wall which absorbs the energy of the airborne sound waves. The heavier and more dense the material of the wall the more effective it is in reducing sound. The Building Regulations 1985 require walls and floors to provide reasonable resistance to airborne sound between dwellings and between machine rooms, tank rooms, refuse chutes and habitable rooms. A solid wall one brick thick, or a solid cavity wall plastered on both sides is generally considered to provide reasonable sound reduction between dwellings at reasonable cost. The small reduction in sound transmission obtained by doubling the thickness of a wall is considered prohibitive in relation to the cost.

For reasonable reduction of airborne sound between dwellings one above the other, a concrete floor is advisable.

The more dense the material the more readily it will transmit impact sound. A knock on a part of a rigid concrete frame may be heard some considerable distance away. Insulation against impact sound will therefore consist of some absorbent material that will act to cushion the impact, such as a carpet on a floor, or serve to interrupt the path of the sound, as for example the absorbent pads under a floating floor.

Noise generated in a room may be reflected from the walls and ceilings and build up to an uncomfortable intensity inside the room particularly where the wall and ceiling surfaces are hard and smooth. To prevent the build-up of reflected sound some absorbent material should be applied to walls and ceilings, such as acoustic tiles or curtains, to absorb the energy of the sound waves.

BRICK AND BLOCK WALLS

The majority of walls of small buildings in this country are built as walls of brick or block. The following is a description of the materials, brick, block and mortar used in the construction of solid and cavity walls.

BRICKS

The word brick is used to describe a small block of burned clay of such size that it can be conveniently held in one hand and is slightly longer than twice its width. Blocks made from sand and lime and blocks made of concrete are manufactured in clay brick size and these are also called bricks. The great majority of bricks in use today are of clay.

The standard brick is $215 \times 102.5 \times 65$ which with a 10 mortar joint becomes $225 \times 112.5 \times 75$ to BS 3921: 1974.

Materials from which bricks are made

Clay: In this country there are very extensive areas of clay soil suitable for brickmaking. Clay differs quite widely in composition from place to place and the clay dug from one part of a field may well be quite different from that dug from another part of the same field. Clay is ground in mills, mixed with water to make it plastic and moulded, either by hand or machine, to the shape and size of a brick.

Bricks that are shaped and pressed by hand in a sanded wood mould and then dried and fired have a sandy texture, are irregular in shape and colour and are used as facing bricks due to the variety of their shape, colour and texture. Machine made bricks are either hydraulically pressed in steel moulds or extruded as a continuous band of clay. The continuous band of clay, the section of which is the length and width of a brick, is cut into bricks by a wire frame. Bricks made this way are called 'wire cuts'. Press moulded bricks generally have a frog or indent

and wire cuts have none. The moulded brick is baked to dry out the water and burned at a high temperature so that part of the clay melts and fuses the whole mass of the brick into a hard durable unit. If the moulded brick is burned at too high a temperature part of the clay melts into a solid glass-like mass and if it is burned at too low a temperature no part of the clay melts and the brick is soft. Neither over-burned nor under-burned brick is satisfactory for building purposes.

A brick wall has very good fire resistance, is a poor insulator against transference of heat, does not, if well built, deteriorate structurally and requires very little maintenance over a long period of time. Bricks are cheap because there is an abundance of the natural material from which they are made, that is clay. The clay can easily be dug out of the ground, it can readily be made plastic for moulding into brick shapes and it can be burned into a hard, durable mass at a temperature which can be achieved with quite primitive equipment.

Because there is wide variation in the composition of the clays suitable for brick making and because it is possible to burn bricks over quite a wide range of temperatures sufficient to fuse the material into a durable mass, a large variety of bricks are produced in this country. The bricks produced which are suitable for building vary in colour from almost dead white to practically black and in texture from almost as smooth as glass to open coarse grained. Some are quite light in weight and others dense and heavy and there is a wide selection of colours, textures and densities between the extremes noted. It is not possible to classify bricks simply as good and bad as some are good for one purpose and not for another. Bricks may be classified in accordance with their uses as commons, facings and engineering bricks or by their quality as internal quality, ordinary quality and special quality. The use and quality classifications roughly coincide, as commons are much used for internal walls, facings or ordinary quality for external walls and engineering or special quality bricks for their density and durability in positions of extreme exposure. In cost, commons are cheaper than facings and facings cheaper than engineering bricks.

Types of bricks

Commons: These are bricks which are sufficiently hard to safely carry the loads normally supported by brickwork, but because they have a dull texture or poor colour they are not in demand for use as facing bricks which show on the outside when built and affect the appearance of buildings. These 'common' bricks are used for internal walls and for rear walls which are not usually exposed to view. Any brick which is sufficiently hard and of reasonably good shape and of moderate price may be used as a 'common' brick. The type of brick most used as a common brick is the Fletton brick.

Facings: This is by far the widest range of bricks as it includes any brick which is sufficiently hard burned to carry normal loads, is capable of withstanding the effects of rain, wind, soot and frost without breaking up and which is thought to have a pleasant appearance. As there are as many different ideas of what is a pleasant looking brick as there are bricks produced, this is a somewhat vague classification.

Engineering bricks: These are bricks which have been made from selected clay, which have been carefully prepared by crushing, have been very heavily moulded and carefully burned so that the finished brick is very solid and hard and is capable of safely carrying much heavier loads than other types of brick. These bricks are mainly used for walls carrying exceptionally heavy loads, for brick piers and general engineering works. The two best known engineering bricks are the red Southwater brick and the blue Staffordshire brick. Both are very hard, dense and do not readily absorb water. The ultimate crushing resistance of engineering bricks is greater than 50 N/mm².

Semi-engineering bricks: These are bricks which whilst harder than most ordinary bricks are not so hard as engineering bricks. It is a very vague classification without much meaning, more particularly as a so-called semi-engineering brick which can safely carry less weight than an engineering brick is not necessarily half the price of an engineering brick.

Composition of clay: Clays suitable for brick making are composed mainly of silica in the form of grains of sand and alumina which is the soft plastic part of clay which readily absorbs water and makes the clay plastic and which melts when burned. Present in all clays are materials other than the two mentioned above such as lime, iron, manganese, sulphur and phosphates. The proportions of these materials varies widely and the following is a description of the composition, nature and uses of some of

the most commonly used bricks classified according to the types of clay from which they are produced.

Flettons: There are extensive areas of what is known as Oxford clay. The clay is composed of just under half silica, or sand, about one-sixth alumina, one-tenth lime and small measures of other materials such as iron, potash and sulphur. The clay lies in thick beds which are economical to excavate. In the clay, in its natural state, is a small amount of mineral oil which, when the bricks are burned, ignites and assists in the burning. Because there are extensive thick beds of this clay, which are economical to excavate, and because it contains some oil, the cheapest of all clay bricks can be produced from it. The name Fletton given to these bricks derives from the name of a suburb of Peterborough around which the clay is extensively dug for brickmaking. Flettons are cheap and many hundreds of millions of them are used in building every year. The bricks are machine-moulded and burned and the finished brick is uniform in shape with sharp square edges or arises. The bricks are dense and hard and have moderately good strength, the average pressure at which these bricks fail, that is crumble, is around 21 N/mm².

The bricks are usually light creamy pink to dull red in colour and because of the smooth face of the brick what are known as 'kiss marks' are quite distinct on the long faces. These 'kiss marks' take the form of three bands of different colours, as illustrated in Fig. 45.

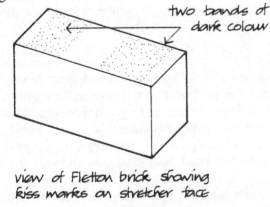

view of Fletton brick showing kiss marks on stretcher face

Fig. 45

The surface is quite hard and smooth and if the brick is to be used for wall surfaces to be plastered, two faces are usually indented with grooves to give the surface a better grip or key for plaster. The bricks are then described as 'keyed Flettons'. Figure 46 is an illustration of a keyed Fletton.

Stocks: The term 'stock brick' is generally used in the south-east counties of England to describe the London stock brick. This is a brick manufactured in Essex and Kent from clay composed of sand and alumina to which some chalk is added. Some combustible material is added to the clay to assist burning. The London stock is usually predominantly yellow after burning with shades of brown, and purple. The manufacturers grade the bricks as 1st Hard, 2nd Hard and Mild, depending on how hard burned they are. The bricks are usually irregular in shape and have a fine sandy texture. Because of their colour they are sometimes called 'yellow stocks'. 1st Hard and 2nd Hard London stocks were much used in and around London as facings as they weather well and were of reasonable price. In other parts of England the term stock bricks describes the stock output of any given brick field.

Marls: By origin the word marl denotes a clay containing a high proportion of lime (calcium carbonate) but by usage the word marl is taken to denote any sandy clay. This derives from the use of sandy clays, containing some lime, as a top dressing to some soils to increase fertility. In most of the counties of England there are sandy clays, known today as marls, which are suitable for brickmaking. Most of the marl clays used for brick making contain little or no lime. Many of the popular facing bricks produced in the Midlands are made from this type of clay and they have a good shape, a rough sandy finish and vary in colour from a very light pink to dark mottled red.

Gaults: The gault clay does in fact contain a high proportion of lime and the burned brick is usually white or pale pink in colour. These bricks are of good shape and texture and make good facing bricks, and are more than averagely strong. The gault clay beds are not extensive in this country and lie around limestone and chalk hills in Sussex and Hampshire.

Clay shale bricks: Some clay beds have been so heavily compressed over centuries by the weight of earth above them that the clay in its natural state is quite firm and has a compressed flaky nature. In the coal-mining districts of this country a considerable quantity of this clay shale has to be dug out to reach coal seams and in those districts the extracted shale is used extensively for brickmaking. The bricks produced from this shale are usually uniform in shape

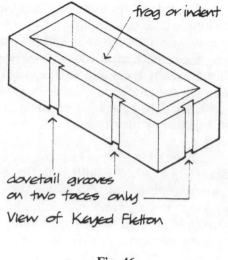

Fig. 46

with smooth faces and the bricks are hard and durable. The colour of the bricks is usually dull buff, grey, brown or red. These bricks are used as facings, commons and semi-engineering, depending on their quality.

Calcium silicate bricks are generally known as sand-lime bricks. The output of these bricks has increased over the past few years principally because the output of Fletton bricks could not keep pace with the demand for a cheap common brick and sand-lime bricks have been mainly used as commons. The bricks are made from a carefully controlled mixture of clean sand and hydrated lime which is mixed together with water, heavily moulded to brick shape and then the moulded brick is hardened in a steam oven. The resulting bricks are very uniform in shape and colour and are normally a dull white. Coloured sand-lime bricks are made by adding a colouring matter during manufacture. These bricks are somewhat more expensive than Flettons and because of their uniformity in shape and colour they are not generally thought of as being a good facing brick. The advantage of them however is that the material from which they are made can be carefully selected and accurately proportioned to ensure a uniform hardness, shape and durability quite impossible with the clay used for most bricks.

Flint-lime bricks: are manufactured from hydrated lime and crushed flint and are moulded and hardened as are sand-lime bricks. They are identical with sand-lime bricks in all respects.

Hollow, perforated and special bricks: Cellular and perforated bricks, illustrated in Fig. 47, are lighter in weight than solid bricks and the cells and perforations facilitate drying and burning. The saving in clay and consequent reduction in weight is an advantage in non-load bearing walls but does not significantly improve thermal insulation in external walls. Cellular bricks are laid with the cells or hollows downwards and perforated bricks should be laid so that the mortar does not fill the perforations.

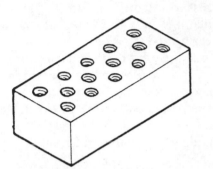

Perforated brick

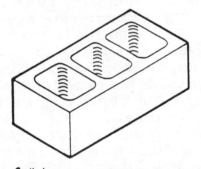

Cellular pressed brick

Fig. 47

The structural 'V' brick developed by the Building Research Station, was designed for use in place of the traditional cavity wall. The vertical perforations provide thermal insulation equivalent to a cavity wall. These bricks can be held in one hand and have to be carefully laid so that mortar does not enter the cavities. The structural 'V' brick is not so easy to handle as the conventional brick as its sharp edges lacerate the hands and the care required in spreading mortar makes it far less popular in use than was originally anticipated.

A range of standard special bricks, illustrated in Fig. 48, is produced for use in facing brickwork for angles, offsets and returns.

Half round coping

Double bullnose

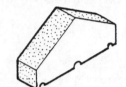
Saddleback coping

Bullnose double stretcher

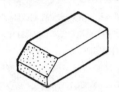
Plinth header

Plinth stretcher

Fig. 48

Properties of bricks

Hardness: This is a somewhat vague term very commonly used in the description of bricks. By general agreement it is recognised that a brick which is to have a moderately good compressive strength, reasonable resistance to saturation by rainwater and sufficient resistance to the disruptive action of frost should be hard burned. Without some experience in the handling, and of the behaviour of bricks in general, it is very difficult to determine whether or not a particular brick is hard burned. A method of testing for hardness is to hold the brick in one hand and give it a light tap with a hammer. The sound caused by the blow should be a dull ringing tone and not a dull thud. Obviously different types of brick will when tapped give off different sorts of sound and a brick that gives off a dull sound when struck may well be hard burned. The student will learn much of the behaviour of a particular type of brick used for different types of wall by examination of existing buildings and questioning those most experienced in the building and design of walls.

Compressive strength: This is the one property of bricks which can be determined accurately. The compressive strength of bricks is found by crushing twelve of them individually until they fail or crumble. The pressure required to crush them is noted and the average compressive strength of the brick is stated as newtons per mm^2 of surface area required to ultimately crush the brick. The crushing resistance varies from about 3.5 N/mm^2 for soft facing bricks up to 140 N/mm^2 for engineering bricks. The following are some comparitive figures:

Mild (i.e. soft) stocks	3.5 N/mm^2
2nd Hard stocks	17.5 N/mm^2
Flettons	21 N/mm^2
Southwater A	70 N/mm^2.

Absorption: Much scientific work has been done to determine the amount of water absorbed by bricks and the rate of absorption, in an attempt to arrive at some scientific basis for grading bricks according to their resistance to the penetration of rain. This work has to date been of little use to those concerned with general building work. A wall built of very hard bricks which absorb little water may well be more readily penetrated by rainwater than one built of bricks which absorb a lot of water. This is because rain will more easily penetrate a small crack in the mortar between bricks if the bricks are dense than if the bricks around the mortar are absorptive. Experimental soaking in water of bricks gives a far from reliable guide to the amount of water they can absorb as air in the pores and minute holes in the brick may prevent total absorption and to find total absorption the bricks have to be boiled in water or heated. The amount of water a brick will absorb is a guide to its density and therefore its strength in resisting crushing but is not a reasonable guide to its ability to weather well in a wall. This term 'weather well' describes the ability of the bricks in a particular situation to suffer rain, frost and wind without losing strength, without crumbling and to keep their colour and texture.

Frost resistance: A few failures of brickwork due to the disruptive action of frost have been reported during the last thirty years and scientific work has sought to determine a brick's resistance to frost failure. Most of the failures reported were in exposed parapet walls or chimney stacks where brickwork suffers most rain saturation and there is a likelihood of damage by frost. Few failures of ordinary brick walls below roof level have been reported. Providing sensible precautions are taken in the design of parapets and stacks above roof level and brick walls in general are protected from saturation by damaged rainwater gutters or blocked rainwater pipes there seems little likelihood of frost damage in this country.

Parapet walls, chimney stacks and garden walls should be built of sound, hard burned bricks protected with coping, cappings and damp-proof courses.

There is a classification of bricks, for frost resistance in BS 3921: 1974 as follows:

(1) Frost resistant (F). Bricks durable in all situations and those repeatedly subject to freezing and thawing.
(2) Moderately frost resistant (M). Bricks durable except when repeatedly subject to freezing and thawing.
(3) Non-frost resistant (O). Bricks liable to damage by freezing and thawing if not protected by cladding. To be used internally.

Efflorescence: Clay bricks contain soluble salts that migrate, in solution in water, to the surface of brickwork as water evaporates to outside air. These salts will collect on the face of brickwork as an efflorescence (flowering) of white crystals that appear in irregular, unsightly patches. This efflorescence of white salts is most pronounced in parapet walls, chimneys and below damp-proof courses where brickwork is most liable to saturation. The concentration of salts depends on the soluble salt content of the bricks and the degree and persistence of saturation of brickwork. The efflorescence of white salts on the surface is generally, merely unsightly and causes no damage. In time these salts may be washed from surfaces by rain. Heavy concentration of salts can cause spalling and powdering of the surface of bricks, particularly those with smooth faces, such as Flettons. This effect is sometimes described as crypto efflorescence. The salts trapped behind the smooth face of bricks, expand when wetted by rain and cause the face of the bricks to crumble and disintegrate.

Efflorescence may also be caused by absorption of soluble salts from a cement rich mortar, concrete behind bricks or from the ground, that appear on the face of brickwork that might not otherwise be subject to efflorescence. Some impermeable coating between concrete and brick can prevent this (see Volume 4). There is no way of preventing the absorption of soluble salts from the ground by brickwork below horizontal d.p.c. level though the effect can be reduced considerably by the use of dense bricks below d.p.c.

Four categories of efflorescence are set out in BS 3921:1974 as:

(1) Nil. No perceptable deposit of salts.
(2) Slight. Up to 10% of the area of the face covered with a deposit of salts, but unaccompanied by powdering or flaking of the surface.
(3) Moderate. More than 10% but not more than 50% of the area of the face covered with a deposit of salts, but unaccompanied by powdering and flaking of the surface.
(4) Heavy. More than 50% of the area of the face covered by a deposit of salts and/or powdering of the surface.

Sulphate attack on mortars and renderings: When brickwork is persistently wet, as in foundations, retaining walls, parapets and chimneys, soluble sulphates in bricks and mortar may in time crystallise and expand and cause mortar and renderings to disintegrate. To minimise this effect bricks with a low sulphate content should be used.

BUILDING BLOCKS

Building blocks are wall units larger in size than a brick that can be handled by one man. Building blocks are made of concrete or clay.

Concrete blocks

These are extensively used for both load bearing and non-load bearing walls, externally and internally. A concrete block wall can be laid in less time and may cost up to half as much as a similar brick wall. Lightweight aggregate concrete blocks have good insulating properties against transfer of heat and have been much used for the inner leaf of cavity walls either with a brick outer leaf or a concrete block outer leaf.

The disadvantage of concrete blocks as a wall unit is that they suffer moisture movement which may cause cracking of applied finishes such as plaster. To minimise cracking due to shrinkage by loss of water, vertical movement joints should be built into long block walls at intervals of up to twice the height of the wall. These movement joints may be either a continuous vertical joint filled with mastic or they may be formed in the bonding of the blocks.

Because the block units are comparatively large any settlement movement in a wall will show more pronounced cracking in mortar joints than is the case with the smaller brick wall unit.

For use as a facing material a wide range of concrete blocks is manufactured from accurate fairface blocks to rugged exposed aggregate finishes.

Concrete blocks are manufactured from cement and either dense or lightweight aggregates as solid, cellular or hollow blocks. A cellular block has one or more holes or cavities that do not pass wholly through the block and a hollow block one in which the holes pass through the block. The thicker blocks are made with cavities or holes to reduce weight and drying shrinkage.

The usual range of size of blocks is 390 long × 190 high × 60, 75, 90, 100, 115, 140, 150, 190 and 200 thick and 440 long × 140, 190, 215 and 290 high × 60, 75, 90, 100, 140, 150, 190, 200, 215, 220 and 225 thick.

Concrete blocks are specified by their minimum average compressive strength for:

(a) all blocks not less than 75 thick and
(b) a maximum average transverse strength for blocks less than 75 thick, which are use for non-load bearing partitions.

The usual compressive strengths for blocks being 2.8, 3.5, 5.0, 7.0, 10.0, 15.0, 20.0 and 35.0 N/mm^2.

Concrete blocks may be classified in accordance with the aggregate used in making the block and the more common uses of the three classes of block as follows:

(1) Dense aggregate concrete blocks for general use in building including below ground: The blocks are made of Portland cement, natural aggregate or blast-furnace slag. The usual mix is 1 part of cement to 6 or 8 of aggregate by volume. These blocks are as heavy per cubic metre as bricks, they are not good thermal insulators and their strength in resisting crushing is less than that of most well-burned bricks. The colour and texture of these blocks is far from attractive and they are usually either painted or covered with a coat of rendering. These blocks are used for internal and external load bearing walls.

(2) Lightweight aggregate concrete blocks for general use in building including below ground, in internal walls and inner leaf of cavity walls: The blocks are made of ordinary Portland cement and one of the following lightweight aggregates: granulated blast-furnace slag, foamed blast-furnace slag, expanded clay or shale, or well-burned furnace clinker. The usual mix is 1 part cement to 6 or 8 of aggregate by volume.

Of the four lightweight aggregates noted, well-burned furnace clinker produces the cheapest block which is about two-thirds the weight of a similar dense aggregate concrete block and is a considerably better thermal insulator. Blocks made from foamed blast-furnace slag are about twice the price of those made from furnace clinker but they are only half the weight of a similar dense aggregate block and have good thermal insulating properties. The furnace clinker blocks are used extensively for walls of houses and the foamed blast-furnace slag blocks for walls of large framed buildings because of their lightness in weight.

(3) Lightweight aggregate concrete blocks primarily for internal non-load bearing walls: The blocks are made with the same lightweight aggregate as those in Class 2. These blocks are more expensive than dense aggregate blocks and are used principally for non-load bearing partitions. These blocks are manufactured as solid, hollow or cellular depending largely on the thickness of the block, the thin blocks being

solid, and the thicker either hollow or cellular to reduce weight and the drying shrinkage of the blocks. Figure 49 is an illustration of typical blocks.

Moisture movement: As water dries out from these precast concrete blocks the shrinkage that occurs, particularly with lightweight blocks, may cause serious cracking of plaster and rendering applied to the surface of a wall built with them. Obviously the wetter the blocks the more they will shrink. It is essential that these blocks be protected on building sites from saturation by rain both when they are stacked on site before use and whilst walls are being built. Clay bricks are small and suffer very little drying shrinkage and therefore do not need to be protected from saturation by rain. Only the edges of these blocks should be wetted to increase their adhesion to mortar when the blocks are being laid.

Clay blocks

Hollow clay building blocks to British Standard 3921: 1974 *Clay bricks and blocks*, are made for use as a wall unit. The blocks are made from selected brick clays that are press moulded and burnt. These hard, dense blocks are hollow to reduce shrinkage during firing and reduce their weight and they are grooved to provide a key for plaster as illustrated in Fig. 50. The standard block is 290 × long × 215 high × 62.5, 75, 100 and 150 thick.

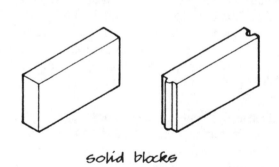

solid blocks

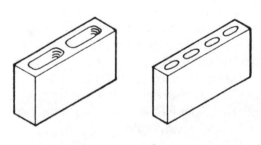

cellular blocks

Fig. 49

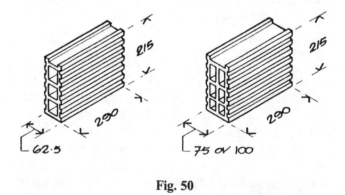

Fig. 50

Clay blocks are comparatively lightweight, do not suffer moisture movement, have good resistance to damage by fire and poor thermal insulating properties. These blocks which are mainly used for non-load bearing partitions in this country are extensively used in southern Europe as infill panel walls to framed buildings.

BONDING

Bonding bricks

In building a wall of bricks or blocks it is usual to lay the bricks in some regular pattern so that each brick bears partly upon two or more bricks below itself. The bricks are said to be bonded, meaning they bind together by being laid across each other. The following is an explanation of the reason for bonding bricks in brickwork.

When a child is first given a set of toy bricks, his natural instinct is to pile them one on another as high as possible, (Fig. 51). He soon finds however that the narrow high stack is very unstable and can easily be knocked down, but, if he arranges them in pyramid fashion, Fig. 52, the stack or wall is much less easily knocked down. If we examine the pyramid style of the child's building we see that the bricks overlap one another, Fig. 53, and this is one kind of bond used in brickwork, namely, 'stretcher' bond. The bonding of bricks in a brick wall is an extension of the use of stone blocks in the pyramids. A wall has to carry the weight of the floors and roofs and we see from the child's experience that a bonded wall more safely carries loads, that is it is less easily over-turned than one not bonded, more particularly as the loads on a wall do not bear down vertically on the centre of the

thickness of a wall. If we look back at the child's first attempt to build we see that his wall consisted of vertical stacks of bricks, next to each other, and that the wall is only as strong as one stack of bricks because if one stack collapses or even one brick is dislodged the whole wall will easily fall down. Because the wall is made of adjacent stacks of bricks there are continuous vertical joints between each stack. It is generally accepted that continuous vertical joints of this sort in a wall of blocks or bricks is a sign of weakness.

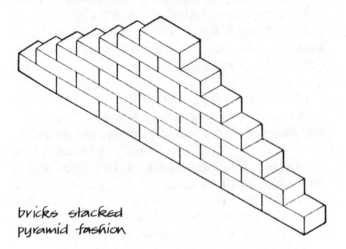

bricks stacked
pyramid fashion

Fig. 52

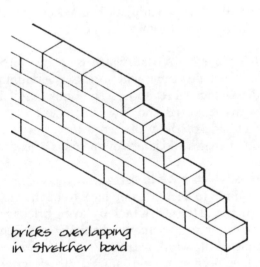

bricks overlapping
in stretcher bond

Fig. 53

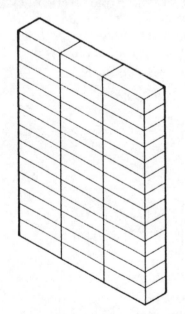

stacks of bricks with continuous
vertical joints between the stacks

Fig. 51

The standard size of a brick is $215 \times 102.5 \times 65$ and for convenience in describing the arrangement of bricks in different bonds the faces of a brick have been named, Fig. 54.

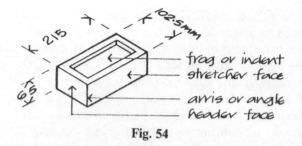

Fig. 54

There are two 'header' faces, each 102.5 mm wide and 65 deep and two 'stretcher' faces each 215 long by 65 deep. The brick is laid on one of its larger 215 × 102.5 mm faces, called the 'bed'. The indent or sinking shown in the sketch is termed a 'frog'. Some bricks have no indent or frog at all, some have one in one face only and some have indents in both long faces.

The indents or frogs in the bed faces of bricks vary in section. Most Flettons have a V-section indent in one face only, and Southwater engineering bricks have an indent roughly rectangular in section. The reason for forming intends in bricks is that a protruding surface on the mould used to give the brick its shape will more surely compact the wet clay than a flat surface. It has been said that the indents or frogs are formed to facilitate the bedding of bricks in mortar but as many bricks, for example those prepared by the wire cut process, have no indent or frog at all and can readily be bedded in mortar, the explanation seems somewhat specious.

Stretcher bond: A wall which is to be 102.5 mm thick is usually built with bricks overlapping or bonded as described above. The bricks are laid on bed with every brick showing a stretcher or long face on each side of the wall, hence the term stretcher bond. At angles and jambs of openings a header face is used.

Walls which have to carry heavy loads or which are required to keep out rain have to be thicker than 102.5 mm and the wall is built by laying bricks side by side to produce any required thickness which is a multiple of the width of a brick.

It has been common practice to describe the thickness of brick walls as 102.5 mm, 215, 327.5 mm, 430 and so on, which presumes that all bricks are the same size. But as bricks vary in size it is more accurate to describe the thickness of walls as ½B, 1B, 1½B, 2B and so on, taking B as being the length of the particular brick used, which in the case of Flettons is 215.

Just as the child's wall was made more secure by overlapping the bricks or bonding them along the length of the wall, so in walls of more than 102.5 mm thickness the bricks should also be bonded or overlapped into the thickness of the wall.

If a wall is to be built 1B thick it is possible to arrange the bricks so that every brick shows a header face on each side of the wall and the bricks along the length of the wall are bonded as shown in Fig 55.

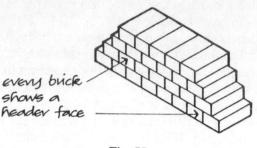

Fig. 55

This arrangement is termed 'heading' or 'header' bond as every brick shows a header face. This type of bond is very little used for 1B thick walls because the general opinion is that the finished effect is not pleasing. The great number of vertical joints between the header faces of bricks usually does not look attractive.

There is only one basic method of arranging or bonding bricks to avoid every brick showing a header face and so that there are no continuous vertical joints, that is to lay the bricks so that the header face of a brick lies directly over the centre of the stretcher face of a brick in the course below and above, Fig. 56.

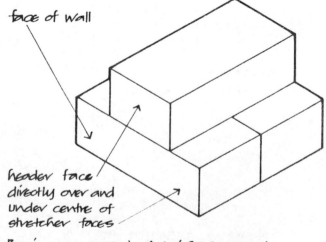

Fig. 56

By trial and error it will quickly be appreciated that apart from header bond this is the only other possible basic arrangement of bonding for walls at least 1*B* thick if the bricks are to be bonded to avoid continuous vertical joints and the instability of walls with such joints.

English and Flemish bond: These are the two types of bond most commonly used and in each the arrangement just described is used, namely a header face centrally over and under a stretcher face.

In English bond the bricks in one course or layer show their header faces and in the course below and above their stretcher faces, as shown in Fig. 57.

In Flemish bond the bricks in every course or layer show alternately header and stretcher faces, Fig. 58.

The header face of many bricks is a darker colour than the stretcher face of the same brick, and it is generally thought that the appearance of a wall in Flemish bond is more attractive than one in English bond. In the former the darker faces are separated by the lighter coloured stretcher faces and in the latter the darker header faces lie in continuous courses.

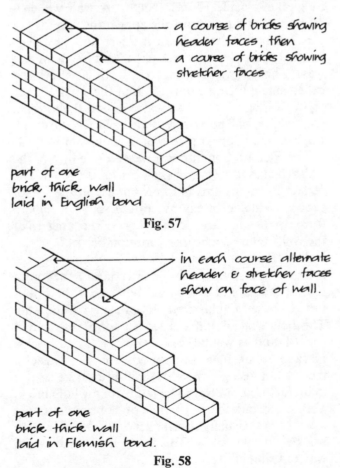

a course of bricks showing header faces, then

a course of bricks showing stretcher faces

part of one brick thick wall laid in English bond

Fig. 57

in each course alternate header & stretcher faces show on face of wall.

part of one brick thick wall laid in Flemish bond.

Fig. 58

Angles and quoins: At the end of a wall where it returns round the corner or angle of the building the arrangement of the bricks described above has to be slightly altered. If we look at the drawing of bonding (Fig. 57) we see that everywhere in the length of the wall bricks project or set back $\frac{1}{4}B$ from the bricks above or below. Obviously if the wall is to end with a straight vertical angle then in some way this $\frac{1}{4}B$ overlap must be filled in or closed. The usual method of filling in this $\frac{1}{4}B$ wide space is to utilise a brick cut in half along its length so that each half of the brick is 1*B* long by $\frac{1}{4}B$ by 65 deep. Because this cut brick is used to fill in or *close* the bond of the bricks it is called a 'closer' and the particular name given to this sized brick is 'queen closer'. It is not always possible to cut a brick cleanly in half and sometimes half queen closers are cut. A full closer and two half closers are shown in Fig. 59.

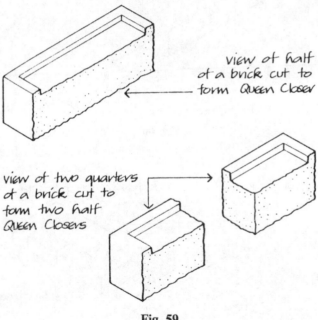

view of half of a brick cut to form Queen Closer

view of two quarters of a brick cut to form two half Queen Closers

Fig. 59

The closer brick is utilised at the angles of walls as in Fig. 60. It will be noticed that the closer brick is not placed at the angle of the wall but $\frac{1}{2}B$ in from the angle next to the header face showing at the angle. The reason for this is that the narrow closer might, if placed at the angle, be easily displaced whilst the wall is being built. The rule for completing or closing the bond at angles is to place a queen closer next to the angle or quoin header in every other course of bricks and this is true irrespective of which sort of bond is used or the thickness of the wall. The word 'quoin' derives from the French word *coin* meaning corner

45

or angle. Figures 61 and 62 show two successive courses of brickwork for walls $1\frac{1}{2}B$ thick in Flemish and English bond. The walls are drawn showing the bonding at angles or quoins.

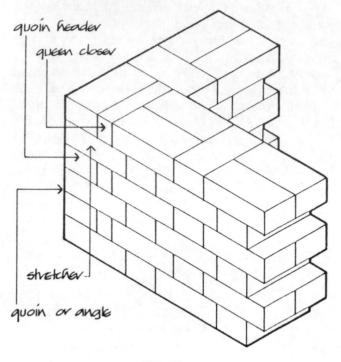

Fig. 60

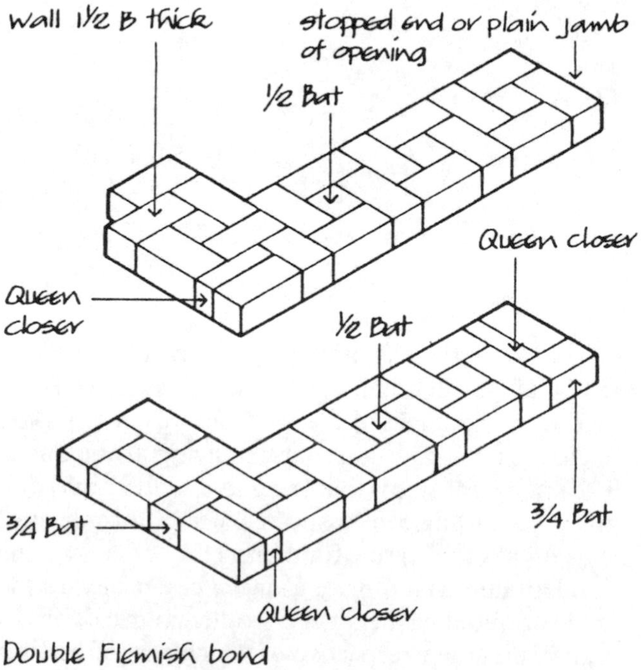

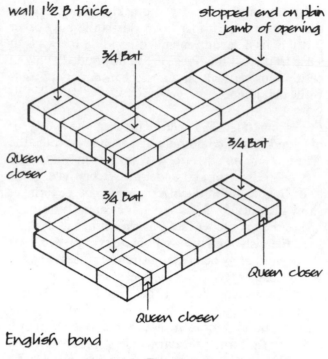

English bond

Fig. 62

Single and double Flemish bond: A wall which is $1\frac{1}{2}B$ thick and is to have the appearance on face of Flemish bond, that is with headers and stretchers alternately in each course, can be built in one of two ways. The first way is to arrange the bricks in what is called single Flemish bond so that the external face shows a Flemish bond appearance and the internal face which will be covered with plaster shows an English bond appearance. This is shown in Fig. 63.

The Flemish bond appearance on one face of the wall is achieved by cutting half bricks as shown. The reason for doing this is to economise in the use of expensive facing bricks by cutting some in half to form two header faces so that most of the inner $1B$ of the wall can be of cheaper common bricks.

The other way to build this wall is to arrange the bricks so that they show Flemish bond appearance on both faces and this would be done if the bricks were to be left unplastered on both sides of the wall. This bond, called double Flemish bond, is shown in Fig. 61. and as will be seen a number of half bricks, marked $\frac{1}{2}$ bat, have to be used. This is a wasteful method of using bricks as half brick size is not manufactured but has to be cut from a whole brick. Very rarely indeed can a whole brick be cut into two exact halves. Usually the smaller 'half' of the broken brick is thrown away. This is obviously a wasteful way of using bricks.

Double Flemish bond

Fig. 61

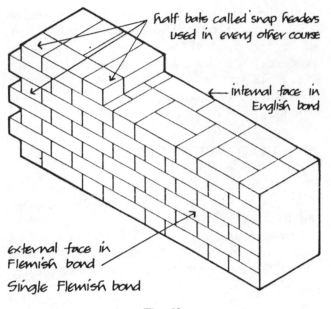

half bats called 'snap headers used in every other course

internal face in English bond

external face in Flemish bond

Single Flemish bond

Fig. 63

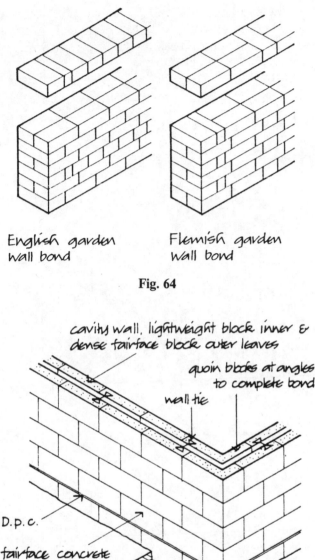

English garden wall bond

Flemish garden wall bond

Fig. 64

cavity wall, lightweight block inner & dense fairface block outer leaves

quoin blocks at angles to complete bond

wall tie

D.p.c.

fairface concrete block wall

trench fill foundation

Concrete block wall

Fig. 65

Garden wall bonds: Walls, such as garden walls, that are to be finished fairface both sides are often built in one of the garden wall bonds. The term, fairface, in relation to brickwork indicates that the wall face is to be exposed with a neat, flush, facing brick finish.

Because of variations in the size and shape of many facing bricks it is difficult to finish a *1B* wall fairface both sides because of differences in the length of bricks that are bonded through the thickness of the wall and show their header faces both sides. Garden wall bonds are designed to reduce the number of header faces to facilitate a fairface finish both sides in walls where appearance is more important than stability.

It will be seen from Fig. 64 that there is one course of header bricks to every three courses of stretchers in English garden wall bond and one header to every three stretchers in each course in Flemish garden wall bond.

Bonding blocks

Concrete blocks are laid in stretcher bond as there is no need to bond into the thickness of the wall. The various thicknesses of block are made to suit most wall thickness requirements. For bonding at angles full blocks are used in running or stretcher bond, where the thickness of the block is half its length as illustrated in Fig. 66. For other thicknesses of block either the off-centre running bond or the thin

stretcher bond are used as illustrated in Fig. 66 with cut closer blocks or ¾ or ⅔ blocks respectively. Quoin blocks are made and used at angles as illustrated in Fig. 65 particularly where a block wall is to be finished fairface. Quoin blocks are little used for other than fairface work as they are liable to damage in handling and use and add considerably to the cost of materials and labour.

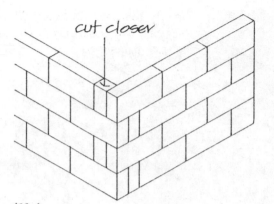

Thin stretcher bond

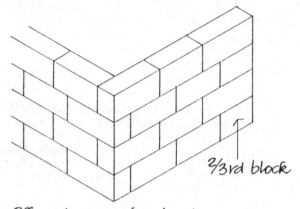

Off centre running bond

Running (stretcher) bond

Bonding building blocks

Fig. 66

At junctions of load bearing concrete block walls it is sometimes considered good practice to butt the end face of one wall to the other with a continuous vertical joint to accommodate shrinkage movements and minimise cracking of plaster finishes. Where one intersecting concrete block wall serves as a buttress to another, the butt joint should be reinforced by building in split end wall ties at each horizontal joint and across the vertical butt joint to bond the intersecting walls. Similarly, non-load bearing concrete block walls should be butt jointed at intersections and the joint may be reinforced with strips of expanded metal in horizontal joints, Fig. 73.

MORTAR FOR BRICKWORK AND BLOCKWORK

Clay bricks are never exactly rectangular in shape and they vary in size. Some facing bricks are far from uniform in shape and size and if a wall were built of bricks laid without mortar and the bricks were bonded the result might be as shown, exaggerated in Fig. 67.

Badly shaped facing bricks laid without mortar

Fig. 67

Because of the variations in shape and size, the courses of bricks would not lie anywhere near horizontal. One of the functions of brickwork is to support floors and if a floor timber were to bear on the brick marked A it would tend to cause it to slide down the slope on which it would be resting. It is essential therefore that brickwork be laid in true horizontal courses, and the only way this can be done with bricks of differing shapes and sizes is to lay them on some material which is sufficiently plastic whilst the bricks are being laid to take up the difference in size, and which must be able to harden to such an extent that it can carry the weight normally carried by brickwork. The material used is termed mortar. The basic requirements of a mortar are that it will harden to such an extent that it can carry the weight normally carried by bricks, without crushing, and that it be sufficiently plastic when laid to take the varying sizes of bricks. It must have a porosity similar to that of the bricks and it must not deteriorate due to the weathering action of rain or frost. Sand is a natural material which is reasonably cheap and which if mixed with water can be made plastic, yet which has very good strength in resisting crushing. Its grains are also virtually impervious to the action of rain and frost. The material required to bind the grains of sand together into a solid mass is termed the matrix and two materials used for this purpose are lime and cement.

Aggregate for mortar

Sand: The aggregate or main part of mortar is sand. The sand is dredged from pits or river beds and a good sand should consist of particles of differing sizes up to 5 in size. In the ground, sand is usually found mixed with some clay earth which coats the particles of sand. If sand mixed with clay is used for mortar, the clay tends to prevent the cement or lime binding the sand particles together and in time the mortar crumbles. It is therefore important that the sand be thoroughly washed so that there is no more than 5% of clay in the sand delivered to site.

Soft sand and sharp sand: Sand which is not washed and which contains a deal of clay in it feels soft and smooth when held in the hand, hence the term soft sand. Sand which is clean feels coarse in the hand, hence the term sharp. These are terms used by craftsmen. When soft sand is used, the mortar is very smooth and plastic and it is much easier to spread and to bed the bricks in than a mortar made of sharp or clean sand. Naturally the bricklayer prefers to use a mortar made with soft or unwashed sand, often called 'builders sand'. A good washed sand for mortar should, if clenched in the hand, leave no trace of yellow clay stains on the palm.

Matrix for mortar

Cement – lime: Cement is made by heating a finely ground mixture of clay and limestone, and water, to a temperature at which the clay and limestone fuse into a clinker. The clinker is ground to a fine powder called cement. The cement most commonly used is ordinary Portland cement which is delivered to the site in 50 kg sacks. When the fine cement powder is mixed with water a chemical action between water and cement takes place and at the completion of this reaction the nature of the cement has so changed that

it binds itself very firmly to most materials. If the cement is thoroughly mixed with sand and water, the reaction takes place, the excess water evaporates leaving the cement and sand to gradually harden into a solid mass. The hardening of the mortar becomes noticeable some few hours after mixing and is complete in a few days. The usual mix of cement and sand for mortar is from 1 part cement to 3 or 4 parts sand to 1 part of cement to 8 parts of sand by volume, mixed with just sufficient water to render the mixture plastic.

A mortar of cement and sand is very durable and is used for all brickwork below ground level and all brickwork exposed to weather above roof level such as parapet walls and chimney stacks.

Cement mortar made with washed sand is not as plastic however as bricklayers would like it to be. Also when used with some types of bricks it can cause an unsightly effect known as efflorescence. This word describes the appearance of an irregular white coating on the face of bricks, caused by minute crystals of water soluble salts in the brick. The salts go into solution in water inside the bricks and when the water evaporates in dry weather they are left on the face of bricks or plaster. Because cement mortar has greater compressive strength than required for most ordinary brickwork and because it is not very plastic, it is not recommended for general brickwork but instead a mixture of lime and cement with sand is used.

Lime: Lime is manufactured by burning limestone or chalk and the result of this burning is a dirty white lumpy material known as quicklime. When this quicklime is mixed with water a chemical change occurs during which heat is generated in the lime and water, and the lime expands to about three times its former bulk. This change is gradual and takes some days to complete, and the quicklime afterwards is said to be slaked, that is it has no more thirst for water. More precisely the lime is said to be hydrated, which means much the same thing. Obviously the quicklime must be slaked before it is used in mortar otherwise the mortar would increase in bulk and squeeze out of the joints. Lime for building is delivered to the site ready slaked and is termed 'hydrated lime'.

When mixed with water, lime combines chemically with carbon dioxide in the air and in undergoing this change it gradually hardens into a solid mass which firmly binds the sand.

A lime mortar is usually mixed with 1 part of lime to 3 parts sand by volume. The mortar is plastic and easy to spread and hardens into a dense mass of good compressive strength. A lime mortar readily absorbs water and in time the effect is to reduce the adhesion of the lime to the sand and the mortar crumbles and falls out of the joints in the brickwork.

Mortar for general brickwork is usually made from a mixture of cement, lime and sand in the proportions set out in table 6. These mixtures combine the strength of cement with the plasticity of lime, have much the same porosity as most bricks and do not cause efflorescence on the face of the brickwork.

The mixes set out in table 6 are tabulated from rich mixes (1) to weak mixes (5). A rich mix of mortar is one in which there is a high proportion of matrix, that is lime or cement or both, to sand as in the 1:3 mix and a weak mix is one in which there is a low proportion of lime or cement to sand as in the mix 1:3:12. The richer the mix of mortar the greater its compressive strength and the weaker the mix the greater the ability of the mortar to accommodate moisture or temperature movements.

Table 6. Mortar Mixes. Proportions by Volume

Mortar designation	Cement:lime: sand	Air-entrained mixes	
		Masonry cement:sand	Cement:sand with plasticiser
1	1:0 to ¼:3		
2	1:½:4 to 4½	1:2½ to 3½	1:3 to 4
3	1:1:5 to 6	1:4 to 5	1:5 to 6
4	1:2:8 to 9	1:5½ to 6½	1:7 to 8
5	1:3:10 to 12	1:6½ to 7	1:8

Taken from BS5628:Part 1:1978

The general uses of the mortar mixes given in table 6 are as mortar for brickwork or blockwork as follows:

Mix 1 For sills, copings and retaining walls
Mix 2 Parapets and chimneys
Mix 3 Walls below d.p.c.
Mix 4 Walls above d.p.c.
Mix 5 Internal walls and lightweight block inner leaf of cavity.

Hydraulic lime is made by burning a mixture of chalk or limestone that contains clay. Hydraulic lime is stronger than ordinary lime and will harden in wet conditions, hence the name. Ordinary Portland cement, made from similar materials and burned at a higher temperature, has largely replaced hydraulic lime which is little used today.

Mortar plasticisers

Liquids known as mortar plasticisers are manufactured. When these liquids are added to water they effervesce, that is the mixture becomes bubbly like soda water. If very small quantities are added to mortar, when it is mixed, the millions of minute bubbles that form, surround the hard sharp particles of sand and so make the mortar plastic and easy to spread. The particular application of these mortar plasticisers is that if they are used with cement mortar they increase its plasticity and there is no need to use lime. It seems that the plasticisers do not adversely affect the hardness and durability of the mortar and they are commonly and successfully used for mortars.

JOINTING AND POINTING

Jointing

Jointing is the word used to describe the finish of the mortar joints between bricks, to provide a neat joint in brickwork that is finished fairface. Fairface describes the finished face of brickwork that will not be subsequently covered with plaster, rendering or other finish.

Most fairface brickwork joints are finished, as the brickwork is raised, in the form of a flush or bucket handle joint. When the mortar has gone off, that is hardened sufficiently, the joint is made. Flush joints are generally made as a 'bagged' or 'bagged in' joint. The joint is made by rubbing coarse sacking or a brush across the face of the brickwork to rub away all protruding mortar and leaving a flush joint. A bucket handle or recessed joint is made by running the top face of a metal bucket handle or the handle of a spoon along the joint to form a concave, recessed joint illustrated in Fig. 68. The advantage of the bucket handle or recessed joint is that the operation compacts the mortar into the joint and improves weather resistance to some extent.

The struck and protruding joints shown in Fig. 68, are more laborious to make and therefore considerably more expensive. The struck joint is made with a

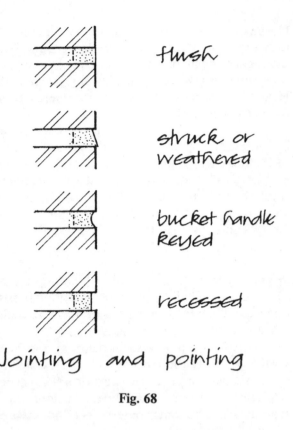

flush

struck or weathered

bucket handle keyed

recessed

Jointing and pointing

Fig. 68

pointing trowel that is run along the joint either along the edges of uniformly shaped bricks or along a wood straightedge, where the bricks are irregular in shape or coarse textured, to form the splayed back joint. The protruding joint is similarly formed with a tool shaped for the purpose, with such filling of the joint as may be necessary to complete the joint.

Pointing

Pointing is the operation of filling mortar joints with a mortar selected for colour and texture to either new brickwork or to old brickwork. The mortar for pointing is a special mix of lime, cement and sand or stone dust chosen to produce a particular effect of colour and texture. The overall appearance of a fairface brick wall can be dramatically altered by the selection of mortar for pointing. The finished colour of the mortar can be affected through the selection of a particular sand or stone dust, the use of pigmented cement, the addition of a pigment and the proportions of the mix of materials.

The joints in new brickwork are raked out about 20 deep when the mortar has gone off sufficiently and before it has set hard and the joints are pointed as

scaffolding is struck, that is taken down.

The mortar joints in old brickwork that was laid in lime mortar may in time crumble and be worn away by the action of wind and rain. To protect the lime mortar behind the face of the joints it is good practice to rake out the perished jointing or pointing and point or repoint all joints. The joints are raked out to a depth of about 20 and pointed with a mortar mix of cement, lime and sand that has roughly the same density as the brickwork. The operation of raking out joints is laborious and messy and the job of filling the joints with mortar for pointing is time consuming so that the cost of pointing old work is expensive.

Pointing or repointing old brickwork is carried out both as protection for the old lime mortar to improve weather resistance and also for appearance sake to improve the look of a wall.

Any one of the joints illustrated in Fig. 68 may be used for pointing.

SOLID AND CAVITY WALLS OF BRICK AND BLOCK

The following is a description of commonly used methods of constructing walls of brick and block to satisfy the functional requirements of small buildings of up to three storeys.

Strength and stability

Up to the middle of this century the design and construction of small buildings, such as houses, was based on tried, traditional forms of construction. There were generally accepted rule of thumb methods for determining the necessary thickness for the walls of small buildings. By and large, the acceptance of tried and tested methods of construction, allied to the experience of local builders using traditional materials in traditional forms of construction, worked well. The advantage was that from a simple set of drawings an experienced builder could give a reasonable estimate of cost and build and complete small buildings, such as houses, without delay.

With the increasing use of unfamiliar materials, such as steel and concrete, in hitherto unused forms from the beginning of this century, it became necessary to make calculations to determine the least size of elements of structure for strength and stability in use. The practicability of constructing large multistorey buildings provoked the need for standards of safety in case of fire and rising expectations of comfort, the need for the control of insulation, ventilation, daylight and hygiene.

During the last fifty years there has been a considerable increase in building control, that initially was the province of local authorities through building by-laws, later replaced by national building regulations. Most recently The Building Regulations 1985 set out functional requirements for buildings that may be met through the practical guidance given in eleven Approved Documents that in turn refer to British Standards and Codes of Practice.

In theory it is only necessary to satisfy the requirements of The Building Regulations 1985 which are short, comparatively simple and include no technical details of means of satisfying the functional requirements. The eleven Approved Documents give practical guidance to meeting the requirements, but there is no obligation to adopt any particular solution in the documents if the requirements can be met in some other way.

The aim of the current Building Regulations is to allow freedom of choice of building form and construction so long as the stated functional requirements are satisfied. In practice the likelihood is that the practical guidance given in the Approved Documents will be accepted as if the guidance were statutory as the easier approach to building, rather than proposing some other form of building that would involve calculation and reference to a bewildering array of British Standards and Codes and Agrément Certificates.

In Approved Document A there is practical guidance to meeting the requirements of Part A of Schedule 1 to The Building Regulations, for loading and ground movement for the walls of small buildings of the following three types:

(a) residential buildings of not more than three storeys,
(b) small single storey non-residential buildings, and
(c) small buildings forming annexes to residential buildings, (including garages and outbuildings)

Limitations as to the size of the building types included in the guidance are given in a disjointed and often confusing manner.

Height: The maximum height of residential buildings is given as 15 m from the lowest ground level to the highest point of any wall or roof, whereas the maximum allowable height of wall is given as 12 m. Height is separately defined, for example, as from the base of a gable end external wall to half the height of the gable. The height of single storey, non-residential buildings is given as 3 m from the ground to the top of the roof, which limits the guidance to very small buildings. The maximum height of an annexe is similarly given as 3 m, yet there is no definition of what is meant by annexe except that it includes garages and outbuildings.

Width: The width of residential buildings is limited to not less than half the height. A diagram limits the dimensions of the wing of a residential building without defining the meaning of the term 'wing', which in the diagram looks more like an annexe than a wing. Whether the arms of a building which is 'L' or 'U' shaped on plan are wings or not is entirely a matter of conjecture. How the dimensions apply to semi-detached buildings or terraces of houses is open to speculation.

In seeking to give practical guidance to meeting functional requirements for strength and stability and at the same time impose limiting dimensions, the Approved Document has created a web of confusion.

One further limitation is that no floor enclosed by structural walls on all sides should exceed 70 m² and a floor without a structural wall on one side, 30 m². The floor referred to is presumably a suspended floor, though it does not say so. As the maximum allowable length of wall between buttressing walls, piers or chimneys is given as 12 m and the maximum span for floors as 6 m, the limitation is in effect a floor some 12×6 m on plan. It is difficult to understand the need for the limitation of floor area for certain 'small' buildings.

Strength: The guidance given in the Approved Document for walls of brick or block is based on compressive strengths of 5 N/mm² for bricks and 2.8 N/mm² for blocks for walls up to two storeys in height, where the storey height is not more than 2.7 m and 7 N/mm² for bricks and blocks of walls of three storey buildings where the storey height is greater than 2.7 m.

Stability: The general limitation of wall thickness given for stability is that solid walls of brick or block should be at least as thick as one sixteenth of the storey height. This is a limiting slenderness ratio relating thickness of wall to height, measured between floors and floor and roof that provide lateral support and give stability up the height of the wall. The minimum thickness of external, compartment and separating walls is given in a table in Approved Document A, relating thickness to height and length of wall as illustrated in Fig. 69. Compartment walls are those that are formed to limit the spread of fire and separating walls (party walls) those that separate adjoining buildings, such as the walls between terraced houses. Cavity walls should have leaves at least 90 thick, cavity at least 50 wide and the combined thickness of the two leaves plus 10, should be at least the thickness required for a solid wall of the same height and length.

Internal load bearing walls, except compartment and separating walls, should be half the thickness of external walls illustrated in Fig. 69 minus 5, except for the wall in the lowest storey of a three storey building which should be of the same thickness or 140, whichever is the greater.

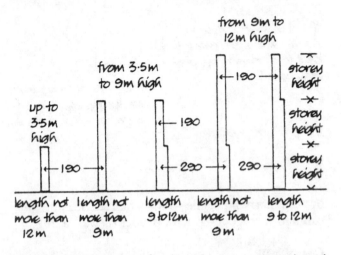

Minimum thickness of external, compartment and separating walls

Fig. 69

Table 7. Lateral Support for Walls

Wall type	Wall length	Lateral support required
Solid or cavity: external compartment separating	Any length	Roof lateral support by every roof forming a junction with the supported wall
	Greater than 3 m	Floor lateral support by every floor forming a junction with the supported wall
Internal load bearing wall (not being a compartment or separating wall)	Any length	Roof or floor lateral support at the top of each storey

Taken from Approved Document A
The Building Regulations 1985

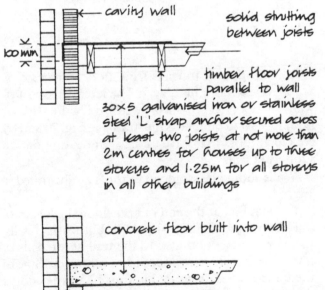

Floors providing lateral restraint to walls

Fig. 70

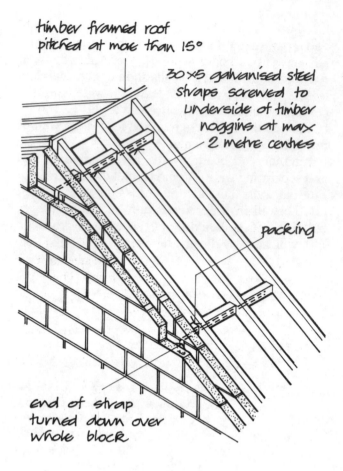

timber framed roof
pitched at more than 15°

30 × 5 galvanised steel
straps screwed to
underside of timber
noggins at max
2 metre centres

packing

end of strap
turned down over
whole block

Lateral support to walls

Fig. 71

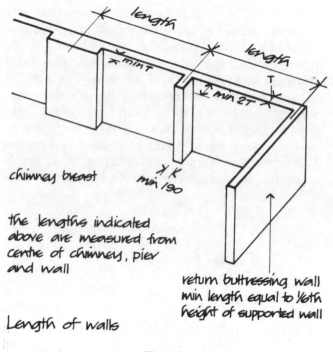

length

length

x min T

x min 2T

T

chimney breast

min 190

the lengths indicated
above are measured from
centre of chimney, pier
and wall

return buttressing wall
min length equal to 1/6th
height of supported wall

Length of walls

Fig. 72

Lateral support to walls from floors and roofs is required as set out in table 7. The lateral support of walls by floors is provided by straps fixed to timber floor joists and built in as illustrated in Fig. 70 and by concrete floors that have at least 90 bearing on the supported wall.

Walls should be strapped to roofs as illustrated in Fig. 71.

For stability at the end of and along the length of load bearing walls there should be buttressing walls, piers or chimneys bonded to the wall at intervals of not more than 12 m measured from the centre line of the buttresses as illustrated in Fig. 72 and extending the full height of the wall.

To limit the effect of chases cut into walls in reducing strength or stability, vertical chases should not be deeper than one-third of the thickness of solid walls or a leaf of a cavity wall and horizontal chases not deeper than one-sixteenth. A chase is a recess, cut

or built in a wall, inside which small service pipes are run and then covered with plaster or wall lining.

Cavity walls

Between 1920 and 1940 it became common for the external walls of small buildings to be constructed as cavity walls with an outer leaf of brick or block, an open cavity and an inner leaf of brick or block. The outer leaf and the cavity serve to resist the penetration of rain to the inside face and the inner leaf to support floors, provide a solid internal wall surface and to some extent act as insulation against transfer of heat.

The idea of forming a vertical cavity in brick walls was first proposed early in the nineteenth century and developed through the century. Various widths of cavity were proposed from the first six inch cavity, a later two inch cavity followed by proposal for three, four or five inch wide cavities. The early cavity walls were first constructed with bonding bricks laid across the cavity at intervals, to tie the two leaves together. Either whole bricks with end closers or bricks specially made to size and shape for the purpose, were used. Later on, during the middle of the century, iron ties were used instead of bond bricks and accepted as being adequate to tie the two leaves of cavity walls.

In the early part of the twentieth century it became normal practice to construct the external walls of houses as a cavity wall with a two inch wide cavity and metal wall ties. It seems that the two inch width of cavity was adopted for the convenience of determining the cavity width, by placing a brick on edge inside the cavity so that the course height of a brick, about 65, determined cavity width rather than any consideration of the width required to resist rain penetration. This was adapted to the two inch (50) wide cavity for walls that became common until recent years.

With the increase in the price of fuels and expectations of thermal comfort, building regulations have of recent years made requirements for the thermal insulation of external walls that can best be meet by the introduction of materials with high thermal resistance. The most convenient position for these lightweight materials in a cavity wall is inside the cavity, which is either fully or partially filled with insulation. A filled or partially filled cavity may well no longer be an efficient barrier to rain penetration so that, with the 1989 increase in requirement for the thermal insulation of walls it has now been accepted that the width of the cavity may be increased from the traditional 50 to 100 to accommodate increased thickness of insulation and still maintain a cavity against rain penetration.

The practical guidance in Approved Document A to The Building Regulations 1985 accepts a cavity of from 50 to 100 for cavity walls having leaves at least 90 thick, built of coursed brickwork or blockwork with wall ties spaced at 450 vertically and from 900 to 750 horizontally for cavities 50 to 100 wide respectively. As the limiting conditions for the thickness of walls related to height and length are the same for a solid bonded wall 190 thick as they are for a cavity wall of two leaves each 90 thick, it is accepted that the wall ties give the same strength and stability to two separate leaves of brickwork that the bond in solid walls does.

Wall ties

Iron ties which were used to tie the leaves of the early cavity walls were later replaced by mild steel ties that became standard for many years. Mild steel suffers progressive deterioration by oxidisation in damp situations, such as external walls and it was considered expedient to coat the standard mild steel ties with zinc to inhibit rust corrosion. The original zinc coating for ties, which was comparatively thin, has been increased in thickness in the current British Standard Specification for improved resistance to corrosion. As added protection, the range of standard wall ties can be coated with plastic on a galvanised undercoating.

In contact with moisture, mild steel progressively corrodes by the formation of oxide of iron, called rust, which expands fiercely to the extent that brickwork around ties may become rust stained and disintegrate.

On the majority of building sites wall ties are not commonly protected during delivery, storage, handling and use against the inevitable knocks that may perforate the toughest coatings to mild steel and the consequent probability of rust occurring. There are, on the market, a range of standard and non-standard section wall ties made from stainless steel, that will not suffer corrosion rusting during the useful life of buildings. It seems worthwhile to make the comparatively small additional expenditure on stainless steel ties as a precaution against staining and spalling of brickwork or blockwork around rusting mild steel ties.

The spacing of wall ties built across the cavity of a

Table 8. Maximum Spacing of Cavity Ties

Width of cavity [mm]	Horizontal spacing [mm]	Vertical spacing [mm]	Other comment
50–75	900	450*	—
76–100	750	450*	Vertical twist type ties† should be used
50–100	—	300	At unbonded jambs to all openings in cavity walls within 150 mm of the opening

Notes
* Or such spacing as will maintain the same number of ties per square metre
† Or ties of equivalent performance

Taken from Approved Document A
The Building Regulations 1985

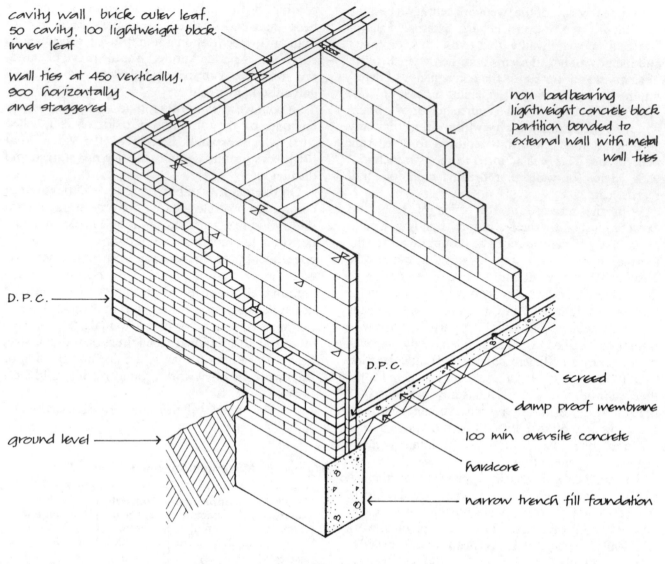

cavity wall, brick outer leaf.
50 cavity. 100 lightweight block
inner leaf

wall ties at 450 vertically.
900 horizontally
and staggered

non loadbearing
lightweight concrete block
partition bonded to
external wall with metal
wall ties

D.P.C.

D.P.C.

ground level

screed

damp proof membrane

100 min oversite concrete

hardcore

narrow trench fill foundation

Fig. 73

cavity wall is usually 900 horizontally and 450 vertically, or 2.47 ties per square metre, and staggered as illustrated in Fig. 73 for the conventional 50 wide cavity, with the spacing reduced to 300 around the sides of openings. In Approved Document A to The Building Regulations, the practical guidance for the spacing of ties is given in table 8.

The standard section wall ties illustrated in Fig. 74 are butterfly and double triangle wire ties and the vertical twist strip tie. As a check to moisture that may pass across the tie, the butterfly type is laid with the twisted wire ends hanging down into the cavity to act as a drip. The double triangle tie may have a bend in the middle of its length and the strip tie has a twist

as a barrier to moisture passing across the tie. Of the three standard types the butterfly tie is more likely to collect mortar droppings then the others.

The wall tie illustrated in Fig. 74 is made from corrosion resistant Austenitic stainless steel. The ridge at the centre of the length of the tie is designed for strength and to provide as small as possible a surface for the collection of mortar droppings. The perforations are to improve bond to mortar.

The length of wall ties varies to accommodate different widths of cavity and the thickness of the leaves of cavity walls. For a 50 cavity with brick leaves, a 191 or 200 long tie is made. For a 100 cavity with brick leaves, a 220 long tie is used.

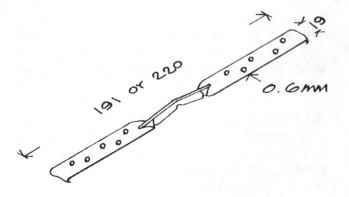

Stainless steel wall tie

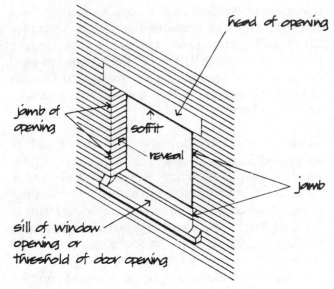

Fig. 75

galvanised steel
Vertical twist wall tie

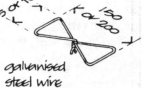

galvanised steel wire
Butterfly wall tie

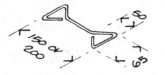

Double triangle wall tie

Cavity wall ties

Fig. 74

OPENINGS IN BRICK AND BLOCK WALLS

The practical guidance in Approved Document A in regard to openings in walls is limited to a statement that the number, size and position of openings should not impair the stability of a wall to the extent that the combined width of openings in a wall between the centre line of buttressing walls or piers should not exceed two-thirds of the length of that wall and a requirement that the bearing end of lintels with a clear span of 1200 or less may be 100 and above that span, 150.

Figure 75 is an illustration of a window opening in a brick wall and the terms used to describe the parts are noted.

For strength and stability the brickwork in the jambs of openings has to be bonded to avoid vertical straight joints and the wall over the head of the opening supported by an arch, lintels or beams.

Jambs of openings

The term jamb derives from the French word *jambe* meaning leg. From Fig. 75, it will be seen that the brickwork on either side of the opening acts like legs which support brickwork over the head of the opening. The term jamb is not used to describe a particular width either side of openings and is merely a general term for the brickwork for full height of opening either side of the window. The word 'reveal' is used more definitely to describe the thickness of the wall revealed by cutting the opening and the reveal is a surface of brickwork as long as the height of the opening. The lower part of the opening is a sill for windows and a threshold for doors.

The jambs of openings may be plain or square as illustrated in Fig. 75, into which the door or window frames are built or fixed or they may be rebated with a recess, behind which the door or window frame is built or fixed.

Plain or square jambs of openings: The opening for windows and doors in solid and cavity brick and block walls may be finished with plain or square jambs. The brickwork in plain or square jambs of openings is bonded with a header face in each alternate course of bricks laid in stretcher bond for cavity walling and with cut blocks to suit the running bonds illustrated in 'Bonding'. Plain jambs in solid brick walls are bonded with a closer next to a header in every other course in the same way that quoins or angles are bonded.

Rebated jambs:
Window and door frames made of soft wood have to be painted for protection from rain, for if wood becomes saturated it swells and in time may decay. With some styles of architecture it is thought best to hide as much of the window frame as possible. So either as a partial protection against rain or for appearance sake, or for both reasons, the jambs of openings are rebated.

Figure 76 shows a view of one rebated jamb on which the terms used are noted.

As one of the purposes of a rebated jamb is to protect the frame from rain the rebate faces into the building and the frame of the window or door is fixed behind the rebate.

Outer reveal:
The brickwork thickness revealed at the sides of the opening on the outside of the rebate is usually $\frac{1}{2}B$ wide for convenience in bonding the bricks. This outer reveal could of course be any width which is a multiple of $\frac{1}{2}B$, providing the wall is thick enough, but usually it is $\frac{1}{2}B$ wide and the width of the *inner* reveal must therefore be the difference between the width of the outer reveal and the thickness of the wall.

Rebate:
The depth of the rebate or recess is usually $\frac{1}{4}B$ but for the cased frames of windows may be $\frac{1}{2}B$ deep. The depth of rebate is made a half of, or the width of, a brick for convenience in bonding the bricks.

Bonding of bricks at jambs:
Just as at an angle or quoin in brickwork, bricks specially cut have to be used to complete, or close, the $\frac{1}{4}B$ overlap caused by bonding, so at jambs special closer bricks $\frac{1}{4}B$ wide on face have to be used.

Provided that the outer reveal is $\frac{1}{2}B$ wide, the following basic rules will apply irrespective of the sort of bond used or the thickness of the wall. If the rebate is $\frac{1}{4}B$ deep the bonding at one jamb will be arranged as illustrated in Fig. 77. In every other course of bricks a header face and then a closer or $\frac{1}{4}B$ wide face must appear at the jamb or angle of opening. To do this and at the same time to form the $\frac{1}{4}B$ deep rebate and to avoid vertical joints continuously up the wall, two cut bricks have to be used.

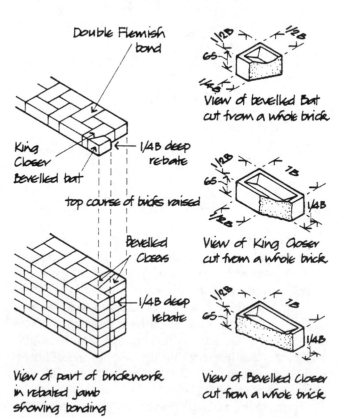

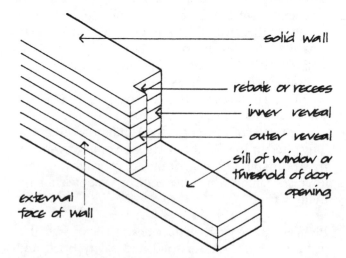

Fig. 76

Fig. 77

These are a bevelled bat (note: a 'bat' is any cut part of a brick) which is shaped as shown in Fig. 77 and a king closer which is illustrated in Fig. 77. Neither of these bricks is made specially to the shape and size shown, but they are cut from whole bricks on the site.

In the course above and below, two other cut bricks, called bevelled closers, should be used behind the stretcher brick. These two bricks are used so as to avoid a vertical joint. Figure 77 shows a view of a bevelled closer.

If the rebate is $\frac{1}{2}B$ deep and the outer reveal is $\frac{1}{2}B$ the bonding at jambs is shown in Fig. 78.

The jambs of openings in walls built of blockwork are seldom rebated because of the difficulty of cutting blocks or using specially shaped blocks.

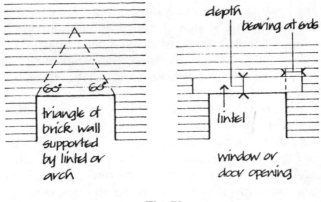

Fig. 79

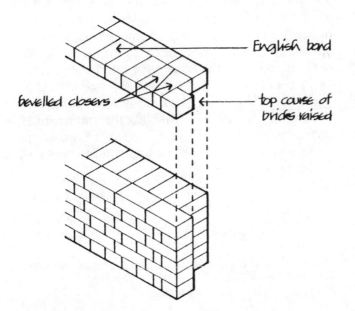

View of part of brickwork in rebated jamb showing bonding

Fig. 78

Head of openings

The brickwork over the head of openings has to be supported either by a flat lintel or an arch. The brickwork which the lintel or arch has to support is a triangle with 60 degree angles, formed by the bonding of bricks, as in Fig. 79. The triangle is formed by the vertical joints between bricks which overlap $\frac{1}{4}B$. In a bonded wall if the brickwork inside the triangle were taken out the rest of the wall above would be quite safe. This is another reason for bonding bricks.

Lintels: This is the name given to any single solid length of concrete, steel, timber or stone built in over an opening to support the wall over it, as shown in Fig. 79. The ends of the lintel must be built into the brick or blockwork over the jambs so as to convey the weight carried by the lintel to the jambs. The area of wall on which the end of a lintel bears is termed its *bearing* at ends. The wider the opening the more weight the lintel has to support and the greater its bearing at ends must be so as to transmit the load it carries to an area capable of supporting it. For convenience in building the ends of a lintel into brickwork its depth is usually made a multiple of brick course height, that is 75, and the lintels are not usually less than 150 deep.

Timber lintels: Up to about sixty years ago it was common practice to support the brickwork over openings on a timber lintel. Wood lintels are not used today because the wood may be damaged during a fire and so cause the collapse of the wall above it and also because timber is liable to rot, again causing the wall over it to collapse.

Concrete lintels: Before the advent of the pressed steel lintel most lintels were of concrete, which was usually reinforced with steel rods. Concrete is a reasonably cheap material, it can easily be moulded or cast when wet and when it hardens it has very good strength in resisting crushing and does not lose strength or otherwise deteriorate when exposed to the weather. The one desirable quality that concrete lacks, if it is to be used as a lintel, is tensile strength, that is strength to resist being pulled apart. The following is an explanation of the behaviour of a lintel when supporting brickwork and the reason for and position of reinforcement in concrete.

Suppose that the lintel over an opening were made of india rubber. When the bricks over the opening are laid the india rubber lintel would bend as shown in Fig. 80. If a piece of india rubber is used as a lintel and is loaded, the top surface becomes squeezed up and the bottom surface, considerably stretched. A squeezing together indicates compression and a stretching tension. A concrete lintel will not bend so obviously as does india rubber under the weight of the brickwork over it, but it will bend very slightly and its top surface will be compressed and its bottom surface stretched or in tension. Concrete is strong in resisting compression but weak in resisting tension, and to give the concrete lintel the strength required to resist the tension at its lower surface, steel is added to it because steel is strong in resisting tension. This is the reason why rods of steel are cast into the bottom of the lintel when it is being moulded in its wet state.

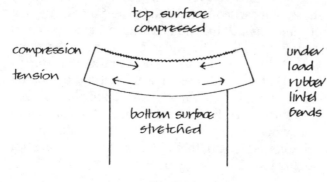

Fig. 80

In fact the india rubber lintel used for the explanation above is not only squeezed up at its top surface and stretched at its bottom surface as can be seen if an india rubber is doubled over. If we examine a folded india rubber we see that it is compressed less and less looking down from the top of it and less and less stretched looking up from the bottom of it towards the centre of the rubber where it is neither stretched nor compressed. In exactly the same way a concrete lintel when carrying brickwork is stretched and compressed most at bottom and top surfaces and less so towards the centre of its depth. Lengths of steel rod are cast into concrete lintels to give them strength in resisting tensile or stretching forces. As

the tension is greatest at the underside of the lintel it would seem sensible to cast the steel rods in the lowest surface. In fact the steel rods are cast in some 15 or more above the bottom surface. The reason for this is that steel very soon rusts when exposed to air and if the steel rods were in the lower surface of the lintel they would rust and in time give way and the lintel would collapse. Also if a fire occurs in the building the steel rods would, if cast in the surface, expand and come away from the concrete and the lintel would collapse. The rods are cast at least 15 up from the bottom of the lintel and the 15 or more of concrete below them is called the concrete cover.

Reinforcing rods: These are usually of round section mild steel of 10 or 12 diameter for lintels up to 1.8 span. The ends of the rods should be bent up at 90° or hooked as in Fig. 81.

The purpose of bending up the ends is to ensure that when the lintel does bend the rods do not lose their adhesion to the concrete around them. After being bent or hooked at the ends the rods should be some 50 or 75 less long than the lintel at either end. An empirical rule for determining the number of 12 rods required for lintels of up to say 1.8 span is to allow one 12 rod for each half brick thickness of wall which the lintel supports.

Fig. 81

Casting lintels: The word 'precast' indicates that a concrete lintel has been cast inside a mould, and has been hardened before it is built into the wall.

The words 'situ-cast' indicate that the lintel is cast in position inside a timber mould fixed over the opening in walls. Whether the lintel is precast or situ-cast will not affect the finished result and which method is used will depend on which is most convenient. It is common practice to precast lintels for most normal door and window openings, the advantage being that immediately the lintel is placed in position over the opening, brickwork can be raised on it whereas the concrete in a situ-cast lintel requires a timber mould or formwork and must be allowed to harden before brickwork can be raised on it.

Lintels are cast in situ, that is in position over openings, if a precast lintel would have been too heavy or cumbersome to have been easily hoisted and bedded in position. Precast lintels must be clearly marked to make certain that they are bedded with the steel reinforcement in its correct place, at the bottom of the lintel. Usually the letter 'T' or the word 'Top' is cut into the top of the concrete lintel whilst it is still wet.

Prestressed concrete lintels: Prestressed precast concrete lintels are used particularly over internal openings. A prestressed lintel is made by casting concrete around high tensile stretched wires which are anchored to the concrete so that the concrete is compressed by the stress in the wires. (See also Volume 4.) Under load the compression of concrete, due to the stressed wires, has to be overcome before the lintel will bend.

Two types of prestressed concrete lintel are made, composite lintels and non-composite lintels.

Composite lintels, which are stressed by a wire or wires at the centre of their depth, are designed to be used with the brickwork they support which acts as a composite part of the lintel in supporting loads. These comparatively thin precast lintels are built in over openings and brickwork is built over them. Prestressed lintels over openings more than 1200 wide should be supported to avoid deflection, until the mortar in the brickwork has set. When used to support blockwork the composite strength of these lintels is considerably less than when used with brickwork.

Non-composite prestressed lintels are made for use where there is insufficient brickwork over to act compositely with the lintel and also where there are heavy loads.

These lintels are made to suit brick and block wall thicknesses as illustrated in Fig. 82. They are mostly used for internal openings, the inner skin of cavity walls and the outer skin where it is covered externally.

Concrete lintels in solid walls: A reinforced concrete or a prestressed lintel may be used over openings in external solid walls with the lintel exposed on the face of the wall or covered by rendering applied to the external face of the wall. Either one or two lintels may be used depending on the convenience of lifting

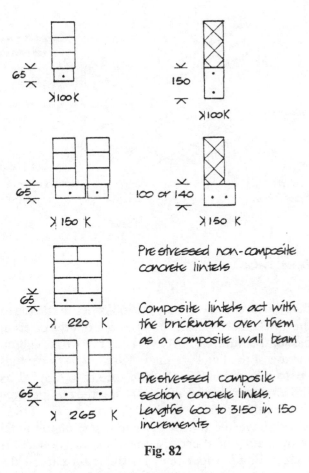

Prestressed non-composite concrete lintels

Composite lintels act with the brickwork over them as a composite wall beam

Prestressed composite section concrete lintels. Lengths 600 to 3150 in 150 increments

Fig. 82

and placing the lintels. The lintel is built into the jambs of the opening sufficient to spread the load. An insulating lining, fixed to the inside face of solid walls, is returned in the soffit of the opening.

Boot lintels: When concrete has dried it is dull light grey colour. Some think that a concrete lintel exposed for its full depth on the external face of brick walls is not attractive. In the past it was common practice to hide the concrete lintel behind a brick arch or brick lintel built over the opening externally. A modification of the ordinary rectangular section lintel, known as a boot lintel, is used to reduce the depth of the lintel exposed externally. Figure 83 shows a section through the head of an opening showing a boot lintel in position. The lintel is boot-shaped in section with the toe part showing externally. The toe is usually made 65 deep and this small depth of concrete showing over the opening does not spoil the appearance of the brickwork. The main body of the lintel is inside the wall where it does not show and it is this part of the lintel which does most of the work of supporting brickwork. Some think

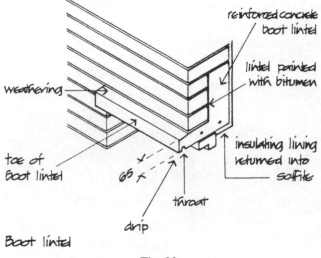

Fig. 83

that the face of the brickwork looks best if the toe of the lintel finishes just 25 or 40 back from the external face of the wall, as in Fig. 84. The brickwork built on the toe of the lintel is usually ½B thick for openings up to 1.8 wide. The 65 deep toe, if reinforced as shown, is capable of safely carrying the two or three courses of ½B thick brickwork over it. The brickwork above the top of the main part of the lintel bears mainly on it because the bricks are bonded. If the opening is wider than 1.8 the main part of the lintel is sometimes made sufficiently thick to support most of the thickness of the wall over, as in Fig. 84. The brickwork resting on the toe of the lintel is built with bricks cut in half. When the toe of the lintel projects beyond the face of the brickwork it should be weathered to throw rainwater out from the wall face and throated to prevent water running in along soffit or underside as shown in Fig. 83. When the external face of brickwork is in direct contact with concrete, as is the brickwork on the toe of this lintel, an efflorescence of salts is liable to appear on the face of the brickwork. This is caused by soluble salts in the concrete being withdrawn when the wall dries out after rain and being left on the face of the brickwork

in the form of white dust. This looks ugly. To prevent it, the faces of the lintel in direct contact with the external brickwork should be painted with bituminous paint. This is indicated in Fig. 83. The bearing at ends where the boot lintel is bedded on the brick jambs should be of the same area as for ordinary lintels.

A boot lintel can be used over openings in a cavity wall only where the wall has an internal insulating lining. Where a cavity wall has an inner skin of lightweight blocks, a cavity fill or a cavity lining a boot lintel used to support both leaves of the cavity would act as a cold bridge through the lintel which has very poor insulating properties as compared to the wall above.

Lintels in cavity walls: Where the thermal insulation in a cavity wall is a cavity fill or partial fill lining applied to the inner leaf, it is important the fill and the lining be carried down to the head of the opening so that the whole of the wall is insulated.

Were the lintel to be built right across the cavity it would act as a cold bridge due to its poor insulating properties and would invite condensation on its inner face. It is, therefore, usual to use two separate lintels one for each of the two skins with the insulation run between them as illustrated in Fig. 85.

Where a lightweight concrete block inner skin is used as the insulation for a cavity wall it is important that either the insulating properties of the blocks be continued down to the head of the opening or that some insulating material be fixed between separate lintels supporting the inner and outer skins. There are on the market galvanised steel lintels specifically designed to support both the outer leaf of a cavity wall and the inner leaf of concrete blocks over the head of openings.

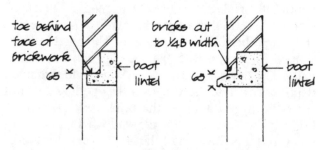

Fig. 84

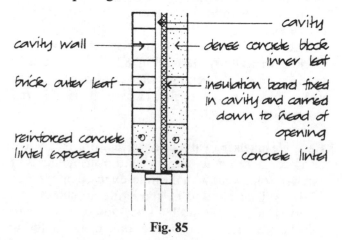

Fig. 85

Steel lintels: Most load bearing brick or blockwork walls over openings are supported by steel section lintels, which are formed by pressing to shapes suited to support flat or curved brick lintels or arches externally and brickwork or blockwork internally for both solid and cavity walls. The advantage of these lintels is that they are comparatively lightweight and easy to handle, they provide adequate support for walling over openings in small buildings and once they are bedded in place the work can proceed without delay. Because of their ease of handling and use these lintels have largely replaced concrete lintels.

The lintels are formed from mild steel strip that is galvanised with a zinc coating to inhibit rust, or from stainless steel. The lintels for use in cavity walling are formed with either a splay to act as an integral damp-proof tray or as a top hat section over which a damp-proof tray is dressed. Typical sections are illustrated in Fig. 86. The splay section lintels are galvanised and coated with epoxy powder coating as corrosion protection and the top hat section with a galvanised

coating. Lintels for openings in solid walls are made of box sections as illustrated in Fig. 86. For internal openings corrugated, tee-section or box section lintels are formed as illustrated in Fig. 88.

For insulation, the top hat section lintels for cavity walling are filled with expanded polystyrene. As key

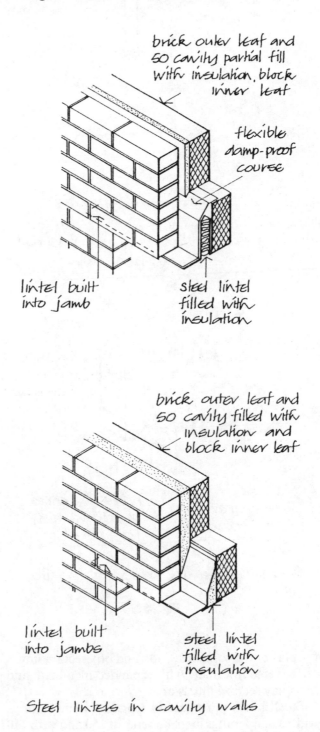

Steel lintels in cavity walls

Fig. 87

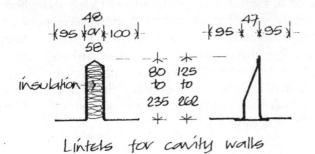

Lintels for cavity walls

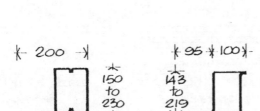

Lintels for solid external walls

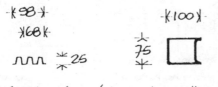

Lintels for internal walls

Fig. 86

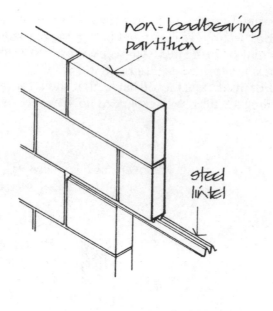

non-loadbearing partition

steel lintel

loadbearing internal partition wall

opening for door

holes in steel lintel as key for plaster

Steel lintels in internal walls

Fig. 88

for plaster the top hat section and box section lintels are covered internally with expanded metal lath and the splay section lintels are perforated.

Steel lintels are made in lengths to suit various widths of openings in increments of 150 and with 150 bearing at each end. A small range of stainless steel lintels is made for external walls.

For openings in cavity walls the splay section lintel supports the outer and inner leaves of the cavity wall and the splay face acts as a damp-proof tray over the head of openings to divert water out through weep holes formed in vertical brick joints as illustrated in Fig. 87. Cavity insulation should be continued down to the top of the lintel and the hollow section filled with insulation to prevent the lintel acting as a thermal bridge.

The top hat section is built in as support for the two leaves of the cavity wall, with a flexible damp-proof tray dressed over the lintel as illustrated in Fig. 87, with the cavity insulation carried down to the top of the lintel.

Steel lintels for internal walls are fixed over openings as illustrated in Fig. 88.

Brick lintels: A brick lintel may be formed as bricks on end, bricks on edge or coursed bricks laid horizontally over openings. The small units of brick, laid in mortar, give poor support to the wall above and usually need some form of additional support. A brick on-end lintel is generally known as a 'soldier arch' or 'brick on-end' arch. The word arch here is wrongly used as the bricks are not arranged in the form of an arch or curve but laid flat. The brick lintel is built with bricks laid on end with stretcher faces showing as in Fig. 89.

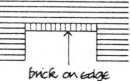

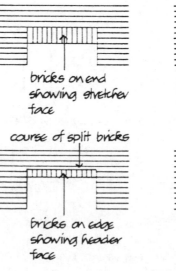

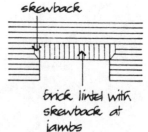

bricks on end showing stretcher face

brick on edge lintel, soffit not in line with courses

course of split bricks

skewback

bricks on edge showing header face

brick lintel with skewback at jambs

Fig. 89

Bricks laid on edge, showing a header face, may be used as a lintel as illustrated in Fig. 89. It will be seen that there has to be a course of split bricks over the lintel. These bricks, cut down to a depth of 37 look untidy. Alternatively the soffit of the lintel can be laid out of line with brick courses (Fig. 89). Only the adhesion between mortar and the bricks keeps these lintels in place and in turn supports the brickwork over the lintel. If the opening over which the brick lintel is formed is wider than say 600 or 900 it is very likely that the adhesion between mortar and bricks in the lintel will weaken in time and one or more of the bricks will sink or even fall out of place. This will obviously look ugly and moreover the bricks above the lintel, which then have no adequate support, may sink. There are various methods of strengthening and giving support to brick lintels. The following are some of them.

For openings up to 900 wide it is generally satisfactory to cut the brick at either end of the lintel on the splay so as to form what is called a skew back, as shown in Fig. 89. It will be seen that the skew or slanting surface bears on a skew brick in the jamb and this improves the stability of the lintel. In building a brick lintel, mortar should be packed tightly between the bricks.

For openings more than 900 wide a brick lintel can be supported by a wrought iron bearing bar, the ends of which are built into brickwork jambs and on which the brick lintel bears (Fig. 90). The bar has bent, or cranked, ends for building into the vertical joints in brickwork at the jambs of the opening. This is not a satisfactory method of supporting the lintel as the bar is visible on the soffit of the lintel and looks ugly, and in time it will rust and look more ugly than it did when first built. Also the bar has a tendency to sag and the lintel may sag with it. As an alternative a mild steel angle section is built with its ends in the brickwork either side of the opening and supporting the lintel. The size of the angle usually used is 50 × 75. This is a satisfactory method of supporting the lintel but unless the door or window frame is fixed near the outside face of the wall the angle will show and in time will rust and look very unsightly.

Another method of support is to drill a hole in each brick of the lintel. This can only successfully be done with fine grained bricks such as Marls or Gaults. Through the holes in the bricks a round-section mild steel rod is threaded and the ends of the rod are built into the brickwork either side of the lintel. This method of supporting the lintel is quite

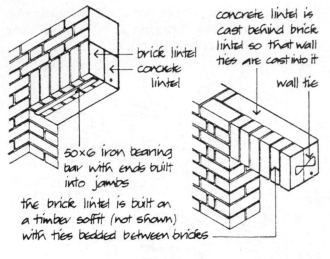

Fig. 90

satisfactory but is somewhat expensive because of the labour involved in drilling the bricks.

A brick lintel can be given support by wall ties bedded between bricks and cast into an in situ cast lintel behind. The lintel is usually built on a timber board called a soffit board which is fixed between the jambs of the opening and is supported from below. As the bricklayer builds the brick lintel he beds a wall tie between the bricks as shown in Fig. 90. When the mortar between bricks has hardened a reinforced concrete lintel is cast in situ behind the brick lintel so that when the concrete lintel has hardened the wall ties are cast into it. This is a sound method of supporting a brick lintel.

In recent years a galvanised pressed steel support for brick lintels has been used. The steel lintel support is shaped for building in over openings and acts as a support for the brick lintel and as a cavity d.p.c. as illustrated in Fig. 87. A galvanised steel lintel is the most commonly used method of supporting brick lintels with the bricks either on end, on edge or coursed across the opening.

Brick arches: Figure 91 shows a semi-circular brick arch on which are noted the various terms used in arch work.

'Intrados' and 'extrados' are the names given to the inside and outside *lines* of curve of an arch. Soffit is the inside curved surface under the arch. The middle third of the arch is called the crown and the two lower thirds, the haunches. Where the plain brickwork meets the extrados of the arch there is an *abutment* of horizontal brickwork with the arch. The horizontal mortar joint or line from which the arch springs is called the springing line. Voussoir is the word used to describe each brick (or stone) used to form an arch.

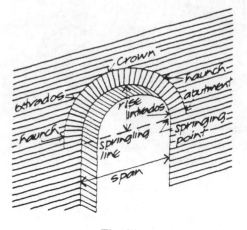

Fig. 91

Rough, axed and gauged bricks: If bricks are used to form a curved arch either the mortar joints between the bricks must be wedge-shaped or the bricks must be cut to a wedge shape and the joints be of uniform thickness, as shown in Fig. 92.

When an arch is built with wedge-shaped joints between bricks which have not been cut, the arch is said to be rough because the wedge-shaped joints look ugly and the finished effect of the arch is rough. Also in time the mortar joints, which may be quite thick at the top of the wedge, will tend to crack and the mortar crumble from the joints. Rough arches are not usually used for facing brickwork.

uncut bricks with wedge shaped mortar joints

bricks cut to a wedge shape & mortar joints of uniform thickness

Fig. 92

Axed bricks: Any good facing brick, no matter how hard, can be cut to a wedge shape on the building site. A template, or pattern, is cut from a sheet of zinc to the exact wedge shape to which the bricks are to be cut. The template is laid on the stretcher or header face of the brick as in Fig. 93. Shallow cuts are made in the face of the brick either side of the template. These cuts are made with a hacksaw blade or file, and

are to guide the bricklayer in cutting the brick. Then, holding the brick in one hand the bricklayer gradually chops the brick to the required wedge shape. For this he uses a tool called a scutch illustrated in Fig. 93. When the brick has been cut to a wedge shape the rough, cut surfaces are made smooth with a coarse rasp which is a steel file with coarse teeth.

From the description this appears to be a laborious operation but in fact the skilled bricklayer can axe a brick to a wedge shape in a few minutes. The axed wedged shaped bricks are built to form the arch with uniform 10 thick mortar joints between the bricks.

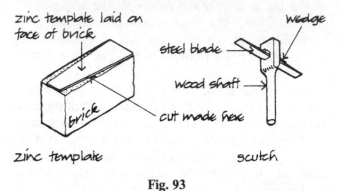

Fig. 93

Gauged bricks: The word gauge means measure, and gauged bricks are those that have been so accurately prepared to a wedge shape that they can be put together to form an arch with very thin joints between them. This does not improve the strength of the brick arch and is done entirely for reasons of appearance. Hard burned clay facing bricks cannot be cut to the accurate wedge shape required for this work because the bricks are too coarse grained, and bricks which are to be gauged are specially prepared. The type of brick used for gauged brickwork is called a rubber brick because its composition is such that it can be rubbed down to an accurate shape on a flat stone.

Rubber bricks are manufactured from fine grained sand and a small proportion of alumina or clay. The bricks are moulded and then baked to harden them, and the temperature at which these bricks are baked is lower than that at which clay bricks are burned, the aim being to avoid fusion of the material of the bricks so that they can easily be cut and accurately rubbed to shape. Rubber bricks have a fine sandy texture and are usually 'brick red' in colour although grey, buff and white rubber bricks are made. These bricks are usually larger than most clay bricks.

Gauged bricks for arch work: The arch to be built is drawn out full size and from this drawing, sheet zinc templates, or patterns, are cut to the exact size of the wedge-shaped brick voussoirs. These templates are placed on the stretcher face or header face of the rubber brick to be cut and the brick is sawn to a wedge shape with a brick saw which consists of an 'H'-shaped wooden frame across which is strung a length of twisted steel wires. Because rubber bricks are soft this twisted wire quickly saws through them.

After the bricks have been cut to a wedge shape they are carefully rubbed down by hand on a large flat stone until they are the exact wedge shape required as indicated by the sheet zinc template.

The gauged rubber bricks are built to form the arch with joints between the bricks usually 1.5 mm thick. A mortar of sand and lime, or cement, is too coarse for so narrow a joint and the mortar used between the gauged bricks is composed of lime and water. The finished effect of accurately gauged red bricks with thin white joints between them is very attractive.

The three most commonly used brick arch shapes are the semi-circular, the segmental and the flat camber arch. There follows a description of each in detail.

Semi-circular brick arch: Rough (uncut), axed or gauged brick voussoirs may be used to form this arch. Which sort of voussoirs is used will depend upon the quality, that is cost, of the brickwork in which the arch is formed. Rough voussoirs are used mainly for large spans such as railway arches supporting a viaduct or for rear elevations of buildings. Axed voussoirs will generally be used for arches in the main elevations of brick walls built of one of the harder coarser grained facing bricks. Gauged voussoirs are generally used for arches in the main elevations of brick walls built of specially selected fine grained facing bricks.

Brick voussoirs, whether they be rough, axed or gauged, can be laid so that either their stretcher or their header face is exposed. Usually semi-circular arches are formed with bricks showing header faces to avoid the excessively broad wedge-shaped voussoirs or joints that occur with stretcher faces showing. This is illustrated by the comparison of two arches of similar span first with stretcher face showing and then with header face showing, Fig. 94. If the span of the arch is of any considerable width, say 1.8 or more, it is often the practice to build it with

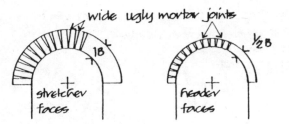
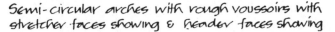
Semi-circular arches with rough voussoirs with stretcher faces showing & header faces showing

Two ring semi-circular arch

Fig. 94

what is termed two or more *rings* of bricks as in Fig. 94. An arch with two rings of bricks is stronger in supporting the weight of the brickwork above than one of only one ring, and generally the wider the span, the more rings. One or more rings may be used whether the bricks are rough, axed or gauged.

Segmental arch: The curve of this arch is a segment, that is part of a circle, and the designer of the building can choose any segment of a circle that he thinks suits his design. By trial and error over many years bricklayers have worked out methods of calculating a segment of a circle related to the span of the arch, which gives a pleasant looking shape of arch, and which at the same time is capable of supporting the weight of brickwork over the arch, such that the rise of the arch is 130 for every metre of span of the arch.

Flat camber arch: This is not a true arch as it is not curved and might well be more correctly named flat brick lintel with voussoirs radiating from a centre (Fig. 95).

The bricks from which the arch is built may either be axed or gauged to the shape required so that the joints between the bricks radiate from a common centre and the widths of voussoirs measured horizontally along the top of the arch are the same. This width will be 65 or slightly less so that there are an odd number of voussoirs, the centre one being a key brick.

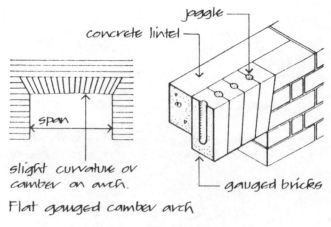

concrete lintel

joggle

span

slight curvature or camber on arch.

Flat gauged camber arch

gauged bricks

Fig. 95

The centre from which the joints between the bricks radiate is usually determined either by making the skew or slanting surface at the end of the arch 60 degrees to the horizontal or by calculating the top of this skew line as lying, 130 for every metre of span of opening, in along the wall over the springing point.

If the underside or soffit of this arch were made absolutely level it would appear to be sagging slightly at its centre. This is an optical illusion and it is allowed for by forming a slight rise or camber on the soffit of the arch. This rise is usually calculated at 6 or 10 for every metre of span and the camber takes the form of a shallow curve.

The camber is allowed for when cutting the bricks to shape. In walls built of hard coarse grained facing bricks this arch is usually built of axed bricks. In walls built of softer fine grained facing bricks the arch is usually of gauged rubber bricks and is termed a flat gauged camber arch. This flat arch must be of such height on face that it bonds in with the brick course of the main walling. The voussoirs of this arch, particularly those at the extreme ends, are often longer overall than a normal brick and the voussoirs have to be formed with two bricks cut to shape.

Flat gauged camber arch: The bricks in this arch are jointed with lime and water, and the joints are usually 1.5 mm thick. As has been said before, lime is soluble in water and does not adhere strongly to bricks as does cement. In time the jointing material, that is lime, between the bricks in this arch may perish and the bricks may slip out of position. To prevent this, joggles are formed between the bricks. These joggles take the form of semi-circular grooves cut in both bed faces of each brick as shown in Fig. 95.

When the bricks are built as an arch there are roughly circular holes between all the bricks and into these holes a wet mix of cement and sand is poured. This hardens and successfully prevents the bricks from slipping out of position. A view of the arch to illustrate the joggles is shown in Fig. 95.

Resistance to weather and ground moisture

The requirements of an external wall to exclude rain depends on the exposure of the wall to wind driven rain. Buildings on high, open ground and those on the sea coast will plainly be more exposed than those that are sheltered in low lying inland towns. The construction and finish of a wall to resist penetration of wind driven rain to its inside face will depend on the severity of its exposure.

British Standard 5628: Part 3 defines six categories of exposure as: very severe; severe; moderate/severe; sheltered/moderate; sheltered; and very sheltered.

Solid walls of brick or block: A solid wall of brick or block will resist the penetration of rain to its inside face by absorbing rainwater that subsequently, in dry periods, evaporates to outside air. The penetration of rainwater into the thickness of a solid wall depends on the exposure of the wall to driving rain and the permeability of the bricks and mortar to water.

The permeability of bricks to water varies widely and depends largely on the density of the brick. Dense engineering bricks absorb rainwater less readily than many of the less dense facing bricks. It would seem logical, therefore, to use dense bricks in the construction of walls to resist rain penetration. In practice, a wall of facing bricks will generally resist the penetration of rainwater better than a wall of dense bricks. The reason for this is that a wall of dense bricks may absorb water through the fine cracks between dense bricks and dense mortar, to a considerable depth of the thickness of a wall and this water will not readily be absorbed through the fine cracks to outside air in dry periods, whereas a wall of less dense bricks and mortar will absorb water to some depth of the thickness of the wall and this water will substantially evaporate to outside air. It is not unknown for a wall of dense bricks and mortar to show an outline of damp stains on its inside face through persistent wetting, corresponding to the mortar joints.

The general rule is that to resist the penetration of rain to its inside face a wall should be constructed of sound, well burned bricks of moderate density, laid in a mortar of similar density and of adequate thickness to prevent the penetration of rain to the inside face.

A solid 1*B* thick wall is not generally sufficiently thick to prevent the penetration of rainwater to its inside face in all but very sheltered positions of exposure. A solid wall 2*B* thick will resist the penetration of rain to its inside face in sheltered positions as set out in table 9, but this thickness of wall is considerably thicker than necessary for strength and stability for small buildings and it does not have good thermal insulating properties. The traditional way of protecting solid walls against rain penetration in severe and very severe positions of exposure is to cover the outside face of the wall with rendering in positions of severe exposure and slate or tile hanging in very severe positions.

To resist the penetration of rain to the inside face, a solid wall of concrete blocks should be of the thicknesses set out in table 9.

Rendering

The word rendering is used in the sense of rendering the coarse texture of a brick or block wall smooth by the application of a wet mix of lime, cement and sand over the face of the wall, to alter the appearance of the wall or improve its resistance to rain penetration, or both. The wet mix is spread over the external wall face in one, two or three coats and finished with either a smooth, coarse or textured finish while wet. The rendering dries and hardens to a decorative or protective coating that varies from dense and smooth to a coarse and open textured.

Stucco is a term, less used than it was, for external plaster or rendering that was applied as a wet mix of lime and sand, in one or two coats and finished with a fine mix of lime or lime and sand, generally in the form imitating stone joints and mouldings formed around projecting brick courses as a background for imitation cornices and other architectural decoration. To protect the comparatively porous lime and sand coating, the surface was usually painted.

The materials and application of the various smooth, textured, rough cast and pebble dash renderings is described in Volume 2.

The materials of an external rendering should have roughly the same density and therefore permeability to water as the material of the wall to which it is applied. There are many instances of the application of a dense rendering to the outside face of a wall that is permeable to water, in the anticipation of protecting the wall from rain penetration. The result is usually a disaster. A dense sand and cement rendering, for example, applied to the face of a wall of porous bricks will, on drying, shrink fiercely, pull away from the brick face or tear off the face of the soft bricks, and the rendering will craze with many fine hair cracks over its surface. Wind driven rain will then penetrate the many hair cracks through which water will be unable to evaporate to outside air

Table 9. Thickness of Single Leaf Walls with or without Rendering

| Exposure category | Minimum thickness of masonry (excluding rendering and finishes) mm | | | | |
| | Clay and calcium silicate masonry | | Concrete masonry | | |
	Rendered	Unrendered	Rendered (dense concrete)	Rendered (lightweight aggregate or autoclaved aerated concrete)	Unrendered
Very severe	Not recommended. Cladding should be used				
Severe	328	Not recommended	250	215	Not recommended
Moderate/severe	215	Not recommended	215	190	Not recommended
Sheltered/moderate	190	440	190	140	440
Sheltered	90	328	90	90	328
Very sheltered	90	190	90	90	190

Taken from BS5390:1976

during dry spells and the consequence is that the wall behind will become more water logged than before and the rendering will have a far from agreeable appearance. The recommended thickness of rendered brick and block wall for various exposure categories is set out in table 9.

Slate and tile hanging

In positions of very severe exposure to wind driven rain, as on high open ground facing the prevailing wind and on the coast facing open sea, it is necessary

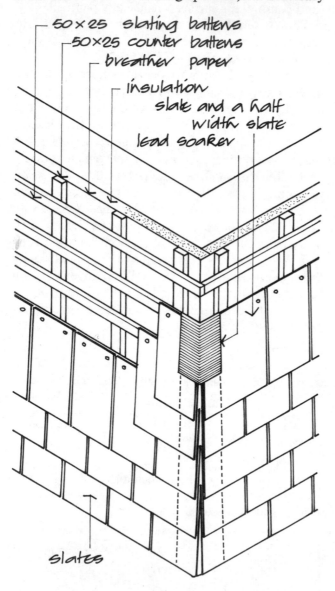

50×25 slating battens
50×25 counter battens
breather paper
insulation
slate and a half
width slate
lead soaker

slates

Slate hanging

Fig. 96

to protect both solid and cavity walls with an external cladding. The traditional form of wall cladding is slate or tile hanging in the form of slates or tiles hung double lap on timber battens nailed to counter battens. Slate hanging has generally been used in the north and tile in the south of Great Britain. Either natural or manufactured slates and tiles can be used.

Pressure impregnated timber battens should be used as a protection against decay and non-ferrous fixings to prevent deterioration due to rusting of steel fixings. Counter battens are fixed up the face of the wall at 300 centres as a fixing for slate and tile battens on which slates and tiles are hung double lap with each slate or tile nailed to the battens with non-ferrous nails. The size of the slates and tiles used are the same as those used for pitched roofs (see Chapter 4).

Where the slate or tile hanging is used as cladding to a solid wall the necessary insulation can be fixed either inside or outside. For outside insulation a water vapour permeable insulation material in the form of rigid boards is fixed to the wall face with the fixings used to secure the counter battens and is covered with a continuous layer of breather paper. Breather paper is resistant to the penetration of water in liquid form but will allow small amounts of water vapour to pass through. The purpose of the breather paper is to protect the outer surface of the insulation from cold air and any rain that might penetrate the hanging and to allow movement of water in the form of vapour through it.

At external and internal angles slates are hung to fit the straight junction of the angles and weathered with sheet lead soakers as illustrated in Fig. 96. Tile hanging may be similarly hung and weathered or special internal or external angle tiles may be used as illustrated in Fig. 97.

Both slate and tile hanging is fixed either to overlap or butt the side of window and door frames which are made and fixed to mask the edges of insulation and battens. The exposed edges of tiles and slates at openings are pointed with cement mortar or weathered with lead flashings.

At the lower edges of slate and tile hanging a projection is formed on or in the wall face by means of blocks, battens or brick corbel courses on to which the lower courses of slates and tiles bell outwards slightly to throw water clear of the wall below.

As an alternative to slate or tile hanging any one of the profiled metal or plastic cladding sheets described in Volume 4 may be used.

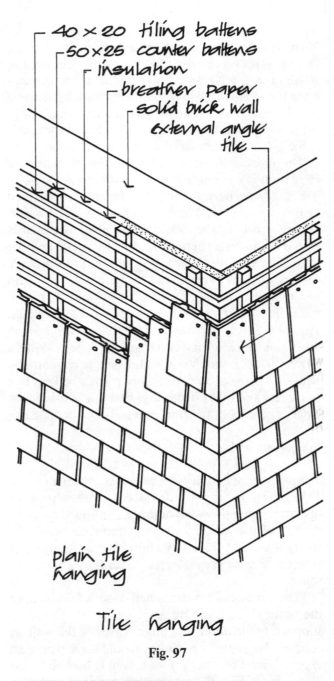

- 40 × 20 tiling battens
- 50 × 25 counter battens
- insulation
- breather paper
- solid brick wall
- external angle tile

plain tile hanging

Tile hanging

Fig. 97

CAVITY WALLS OF BRICK AND BLOCK

Cavity walls were first used early in the nineteenth century. At that time many bricks were fired (burnt) in crude clamps that consisted of moulded bricks that were laid in the form of a low rectangular structure, through which flueways and fire boxes were formed around the bricks laid on edge and spaced apart. The size of the clamp was limited by the ability of the moulded bricks to support the weight of the layers of bricks above, without undue deformation.

A fire was lit in a recess in one face of the clamp, depending on wind direction, so that the fire spread along the flueways to ignite fuel in fire boxes in the heart of the clamp. Depending on the weather, and skill in forming flueways and fire boxes throughout the clamp, the bricks gradually heated, dried and burned over the course of several days. Because of the comparatively crude construction of these clamps, the drying and burning of bricks was far from uniform throughout the clamp. Those bricks around the fires and flueways often burned to a hard, black solid finish and those on the outsides of the clamp were often poorly burned and soft. In consequence, the density and porosity of clamp burned bricks varied considerably. In a wall, the least burned bricks laid as headers into the thickness of a solid 1B wall, would often allow penetration of rain to the inside face. To prevent this, the cavity wall was first introduced and later on metal wall ties were used across the cavity instead of the original brick ties, to prevent the penetration of rain to the inside face.

The width of the cavity does not affect the ability of a cavity wall to resist rain penetration, a continuous, completely clear 25 wide cavity would be just as effective as one 100 wide. The conventional 50 (2 inch) wide cavity was adopted for the convenience of laying the two leaves of the wall. At one time it was considered advisable to ventilate the cavity by encouraging air movement inside the cavity through weep holes in the outer leaf to avoid stagnant moist air in the cavity making the inner leaf damp. Were cold outside air to enter the cavity, it would make the inside leaf cold and so reduce the insulating effect of still air in the cavity. The idea of ventilating the cavity was, therefore, abandoned. Recently, with the introduction of heavy inside insulation it has been reintroduced to prevent vapour pressure condensation on the outside face of vapour barriers to inside insulation.

The important consideration in maintaining the cavity as an effective barrier to the penetration of rain, is to keep the cavity free from mortar droppings, offcuts of brick or block, or anything else that might act to allow rain to penetrate across the cavity to the inner leaf and also to solidly fill both horizontal and vertical joints in the outer leaf, with mortar. Cavity wall ties are shaped to provide a difficult path for water to run across them to the inner leaf. As an additional precaution against water running across, it is recommended that ties be bedded across the two leaves of the wall so that they slope down slightly towards the outside face.

The most effective way of keeping the cavity free of mortar droppings and other debris is to suspend a batten of wood, wrapped in sacking, inside the cavity as the wall is built and to take the batten out and clean it from time to time. A convenient time to do this is at the level where wall ties have to be built in. To remove such mortar droppings that may have fallen into the cavity, it is sometimes good practice to lay pockets of bricks in dry sand at the base of the wall. Once the wall is completed the bricks in sand are removed, loose droppings broken up and raked and flushed out with water. The loose bricks are then bedded in place in mortar. Obviously, this can only be effected where the cavity ends at or above ground or floor slab level. Bricklayers are usually very unwilling to suspend battens inside a cavity, as the raising and cleaning of the battens impedes the rapid completion of the work and affects their pay.

In laying bricks, it is common practice to 'butter' the edges of each brick that will form the sides of exposed vertical joints in fairface work, by forming a small wedge-shaped blob of wet mortar on each brick as it is laid. This has the effect of giving the appearance on the external face that the vertical joints have been filled with mortar, whereas in fact, there is a comparatively thin blob of mortar through which rainwater may penetrate. For resistance to rain penetration it is of importance that vertical joints be solidly filled with mortar. This is a time consuming operation that bricklayers are generally most unwilling to undertake.

In all but positions of very severe and severe exposure, an outer leaf of brick $\frac{1}{2}B$ thick, built with sound, well burned bricks laid in mortar of roughly the same density as the bricks, with all joints solidly flushed up with mortar, will be effective in resisting the penetration of rainwater to the inside face of a cavity wall.

A dense aggregate concrete block outer leaf of similar thickness, in mortar of roughly the same density, with all joints flushed up with mortar, will serve to resist the penetration of rain to the inside face of a cavity wall.

OPENINGS IN CAVITY WALLS

With the 1989 revision of Part L of Schedule 1 of The Building Regulations 1985 and Approved Document L, to require a maximum U value of 0.45 for walls, it will be necessary to use an insulating material at least 30 thick either in the cavity or on the inside or outside face of walls. As concrete blocks with a compressive strength of 2.8 N/mm^2, as required in the Regulations for strength, would, by themselves, have to be inordinately thick to provide the required insulation, the main insulation will have to be through the use of some lightweight material with a low U value.

The most convenient place for this thickness of insulation is inside the cavity, either as cavity fill or preferably as partial fill with the insulation fixed against the inner leaf and a clear cavity of 50 width. Where most of the readily available, reasonably priced insulating materials are used, it will be necessary to use a thickness of about 50 to provide the new U value in most wall constructions. It would be possible to form a cavity 75 wide and fix the 50 thick insulation, leaving a cavity of 25. It would require more care in construction than is common, to keep this narrow cavity completely free. With a cavity 100 wide and 50 of insulation the resultant 50 wide air space cavity could more readily be kept free of mortar and other droppings that might bridge the cavity and encourage rain penetration.

Lintels and arches

The purpose of a cavity in a wall is as a barrier to rain penetration to the inside face. At the head of openings either the cavity must be continued down to the top of the window or door frame or where the cavity is wholly bridged by a lintel or arch, there must be some form of vertical d.p.c. to serve as a barrier to rain penetration.

Where the cavity is continued down to the head of the frame there must be some means of separately supporting the outer and inner leaves of the wall. A concrete lintel under each leaf would serve where an exposed lintel on the face of a wall is acceptable as illustrated in Fig. 98 or if the outside of the wall were to be covered with rendering or other coating. Where the outside leaf is finished as fairface brickwork some form of arch could be used, or more usually a brick lintel. Brick lintels need some form of support, either from galvanised steel or stainless steel angles or one of the purpose-made galvanised, pressed steel lintels. It will be seen from Fig. 99, that the steel lintel is designed to support both the outer and inner leaves of the wall over openings and is shaped as a cavity tray to direct any water that might collect in the cavity, out through weep holes in vertical mortar joints. Where the cavity width is increased to 100 a special width of steel lintel will be necessary. Where a hollow-shaped steel lintel interrupts the cavity insulation, it is necesarry to fill the hollow lintel with an insulating material to avoid a cold bridge.

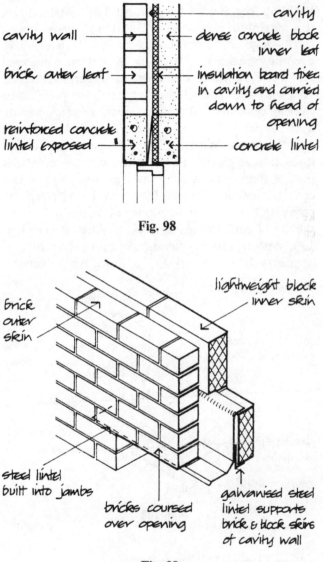

Fig. 98

Fig. 99

Cavity tray

In positions of severe to moderate and moderate exposure to wind driven rain, the outer leaf of a cavity wall will absorb water to the extent that rainwater may penetrate to the cavity side of the outer leaf. It is unlikely that water will enter the cavity unless there are faults in construction. Where the mortar joints in the outer leaf of a cavity wall are inadequately flushed up with mortar, or the bricks in the outer leaf are grossly porous and where the wall is subject to severe or very severe exposure, there is a likelihood that wind driven rain will penetrate through the outer leaf to the cavity. Where rain penetrates the outer leaf and enters the cavity, it is certain that either the construction of the wall is poorly executed, or the bricks have been unwisely chosen or the outer leaf is of inadequate thickness for the position of exposure in which it is built. The solution is to choose some alternative form of construction more suited to the position of exposure.

It is often considered good practice to use some form of impermeable cavity tray above all horizontal breaks in a cavity wall against the possibility of rainwater penetrating the outer leaf. The profile of the pressed steel lintel illustrated in Fig. 99, serves this purpose. When the two leaves of the wall are separately supported and the cavity is continued down as illustrated in Fig. 98, it is practice to build in some form of cavity tray as shown. Strips of polymer-based polythene, bitumen felt or sheet lead may be used for the purpose. The thinner and more flexible polymer-based polythene or lead sheets are preferred rather than bitumen felt. To accommodate a reasonable width of sheet for the tray, it may be necessary to cut the insulation boards in the cavity so that the tray can be built into the inner leaf. The tray should extend at least 150 each side of the opening to shed any water that might collect, clear of the opening. Weep holes are formed in the outer leaf above the lower edge of the tray, where it is built into the outer leaf, by raking out some of the wet mortar from two or more vertical joints in the brickwork.

Jambs of openings

The cavity in a cavity wall serves to prevent penetration of water to the inner leaf. In the construction of the conventional cavity wall before the adoption of cavity insulation, it was considered wise to close the cavity at the jambs of openings to maintain comparatively still air in the cavity as insulation. It was practice to build in cut bricks or blocks as cavity closers. To prevent penetration of water through the solid closing of cavity walls at jambs, a vertical d.p.c. was built in as illustrated in Fig. 100. Strips of bitumen felt or lead were nailed to the back of wood frames and bedded between the solid filling and the outer leaf as shown.

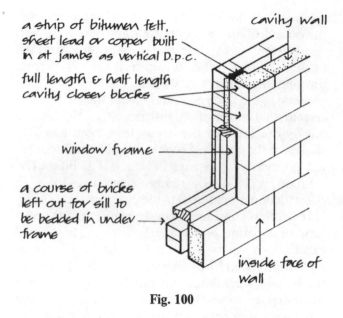

a strip of bitumen felt, sheet lead or copper built in at jambs as vertical D.p.c.

full length & half length cavity closer blocks

window frame

a course of bricks left out for sill to be bedded in under frame

cavity wall

inside face of wall

Fig. 100

As an alternative to solidly filling the cavity at jambs with cavity closers, it was practice to use window or door frames to cover and seal the cavity. Pressed metal subframes to windows were specifically designed for this purpose as illustrated in Fig. 101. With mastic pointing between the metal subframe and the outer reveal, this is a satisfactory way of sealing cavities.

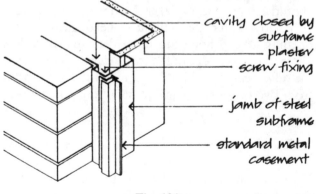

cavity closed by subframe

plaster

screw fixing

jamb of steel subframe

standard metal casement

Fig. 101

At the time when it first became common practice to solidly fill the cavity at jambs, there was no requirement for the insulation of walls. When insulation became a requirement it was met by the use of lightweight, insulating concrete blocks as the inner leaf and the practice of solid filling of cavity at jambs, with a vertical d.p.c. continued. With the increasing requirement for insulation it has become practice to use cavity insulation as the most practical position for a layer of lightweight material. If the cavity

insulation is to be effective for the whole of the wall it must be continued up to the back of window and door frames, as a solid filling of cavity at jambs would be a less effective insulator and act as a thermal or cold bridge.

With the revision of the requirements of The Building Regulations for enhanced insulation down to a maximum U value of 0.45 and more pronounced cold bridge effect at solid filling, it will most likely become practice to use cavity insulation continued up to the frames of openings, as illustrated in Fig. 102, with door and window frames set in position to overlap the outer leaf with a resilient mastic pointing as a barrier to rain penetration between the frame and the jamb. With a cavity 100 wide and cavity insulation as partial fill, it is necessary to cover that part of the cavity at jambs of openings, that is not covered by the frame. This can be effected by covering the cavity with plaster on metal lath or by the use of jamb linings of wood as illustrated in Fig. 103. With this form of construction at the jambs of openings there is no advantage in forming a vertical d.p.c. at jambs.

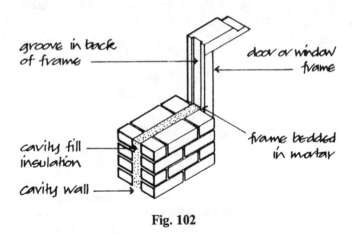

groove in back of frame

door or window frame

cavity fill insulation

cavity wall

frame bedded in mortar

Fig. 102

Sills and thresholds of openings

A sill is the horizontal finish to the wall below the lower edge of a window opening on to which wind driven rain will run from the hard, smooth, impermeable surface of window glass. The function of a sill is to protect the wall below a window. Sills are formed below the sill of a window and shaped or formed to slope out and project beyond the external face of the wall, so that water runs off. The sill should project at least 45 beyond the face of the wall below and have a drip on the underside of the projection.

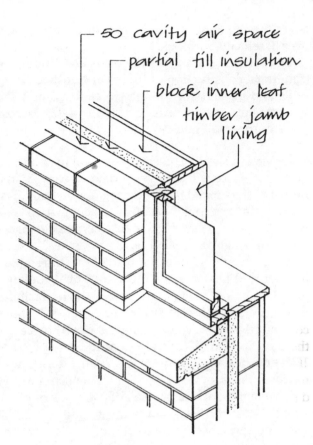

- 50 cavity air space
- partial fill insulation
- block inner leaf
- timber jamb lining

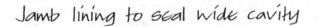

Jamb lining to seal wide cavity

Fig. 103

A variety of materials may be used as a sill such as natural stone, cast stone, concrete, tile, brick and non-ferrous metals. The choice of a particular material for a sill depends on cost, availability and to a large extent on appearance. Details of the materials used and the construction of sills are given in Chapter 2, Volume 2.

As a barrier to the penetration of rain to the inside face of a cavity wall it is good practice to continue the cavity up and behind sills as illustrated in Volume 2. Where sills of stone, cast stone and concrete are used the sill may extend across the cavity. As a barrier to rain penetration it has been practice to bed a d.p.c. below these sills and extend it up behind the sill as illustrated in Volume 2. Providing the sill has no joints in its length, its ends are built in at jambs and the material of the sill is sufficiently dense to cause most of the rainwater to run off, there seems little purpose in these under sill d.p.c.'s or trays.

The threshold to door openings serves as a finish to protect a wall or concrete floor slab below the door as illustrated in Chapter 2 of Volume 2. Thresholds are commonly formed as part of a step up to external doors as part of the concrete floor slab with the top surface of the threshold sloping out. Alternatively a natural stone or cast stone threshold may be formed.

Durability and freedom from maintenance

A wall built with sound, well burned bricks laid in a mortar of roughly the same density as the bricks and designed with due regard to the exposure of the wall to driving rain and with sensible provisions of damp-proof courses to walls around openings and to parapets and chimneys, should be durable for the anticipated life of the majority of buildings and require little if any maintenance and repair. In time, these materials exposed to wind and rain will slowly change colour. This imperceptible change will take place over many years and is described as weathering, that is a change of colour due to exposure to weather. It is generally accepted that this change, due to weathering, enhances the appearance of brick and stone walls.

Walls built of brick laid in lime mortar may in time need repointing, to protect the mortar joints and maintain resistance to rain penetration and to improve the appearance of the wall.

The surface of rendering applied to walls, will in time become weather stained by particles of dust carried in rainwater that runs down the face of the rendering. The appearance of coarse textured renderings and pebble dash (see Volume 2) show the effect of irregular stains less than smooth rendering and the fine drying shrinkage cracks common with rich mixes of rendering, are less obvious on coarse textures. For appearance sake, rendered finishes are often painted. As the function of the rendering is to improve the resistance to rain penetration it is sensible to apply a painted coating that will not impede drying out during dry spells. There are exterior quality 'emulsion' paints that will successfully cover the surface below, for appearance sake, yet will to an extent allow evaporation of water through their surface. These coatings are said to 'breathe'.

Concrete blocks are little used in this country as a fairface finish to walls because the dull, metallic grey colour of the material does not look well in this cold, grey, northern climate. Smooth faced blocks tend to irregular staining by rain, particularly at the run-off of water at sills and other projections, and the mortar joints may stain to a different extent from the blocks around them. It is generally accepted that staining is not an attractive feature of fairface concrete blocks.

Concrete block walls are generally covered with rendering for appearance sake and the rendering behaves in the same way that it does on a brick backing except that there may be more pronounced cracking around mortar joints. Similarly, clay building blocks, which are very commonly used in southern Europe, are usually covered with rendering which does not show cracking at mortar joints due to the very small moisture or temperature movement of these blocks.

Fire resistance

The requirements for the maintenance of structural stability and limitation of spread of fire within and between buildings for houses, can be met by traditional construction of brick or block walls, timber boarded floors and timber roofs covered with slates or tiles. Means of escape requirements do not apply to two storey houses.

The minimum period of fire resistance for houses is set out in table 1.1 of Approved Document B2/3/4 to The Building Regulations in which the minimum period of fire resistance for houses of up to three storeys is given as 30 minutes for all structural elements above ground and one hour for basements 50 m² or more in area and one hour for separating walls.

Table A3 of Appendix A of Approved Document B notes periods of fire resistance for particular elements of construction. A masonry wall, with or without plaster finish, at least 90 thick (75 if non-load bearing) has a notional resistance of one hour which is greater than the minimum period required for houses in Table 1.1. The notional resistance of a timber floor with joists at least 37 wide, covered with at least 15 t & g boarding or sheets of plywood or wood chipboard and a ceiling of 12.5 mm plasterboard with at least 5 neat gypsum plaster finish is 30 minutes, or with 30 plasterboard in two layers with joints staggered and exposed joints taped and filled, one hour. These two forms of floor construction will

cover the requirements for upper floors and floors over basements for houses.

The internal surfaces of the walls and ceilings of traditional construction of brickwork or blockwork walls and plasterboard linings to ceilings meet the requirements for control of spread of fire across surfaces for small houses.

Concealed spaces: The cavity in a cavity wall is a concealed space which might provide a path for the spread of smoke and flames during fires in buildings. A requirement from Part B of Schedule 1 to The Building Regulations 1985 is that concealed spaces (cavities) shall be sealed and subdivided where necessary to inhibit unseen spread of fire and smoke.

Approved Document B giving practical guidance to meeting the requirements of Part B, includes Appendix G, concealed spaces. section G3(1) states that the edges of cavities should be closed by cavity barriers. Section G3(2) states that the provisions of G3(1) do not apply to a cavity in an external wall shown in diagram G2. Diagram G2, which is a very simple diagram of a cavity wall of two brick or block leaves at least 75 thick and a cavity not more than 100 wide, indicates that the cavity should be closed at the top of the wall and at the top of openings.

In the notes to this diagram is a list of materials that may be placed in or exposed to the cavity, such as timber lintel, window or door frame, ends of timber joists, d.p.c, wall ties and thermal insulating material which (unless it is in a house) fills the cavity. The implication of these tortuous notes in relation to thermal insulating material is that combustible material may not be used for the total fill to the cavity in a masonry wall to a house. In other words only non-combustible material such as glass fibre or rockwool can be used as total fill to the cavity of an external wall of a house. In which case the incombustible fill will also serve to close the cavity at the top of the wall and the top of openings.

As partial fill with combustible insulating material is not excluded, the combustible organic materials EPS, XPS, PIR and PUR may be used providing the cavity is closed at the top of the wall and the top of openings.

In Section G8 of Approved Document B the list of materials that may be used as cavity barriers include steel at least 3 thick, timber at least 38 thick, wire reinforced mineral wool blanket at least 50 thick and material of limited combustibility such as cement mortar at least 25 thick. With partial fill insulation to a cavity, the guidance in Approved Document B would be met by the use of wire reinforced mineral wool blanket, pressed into the top of the cavity or a capping of brick or block and at the head of openings a pressed steel lintel or mineral wool blanket as cavity barriers.

Resistance to the passage of heat

To maintain reasonable conditions of thermal comfort in buildings and to economise in the use of fuel, it is accepted that the exposed parts of a building should provide a minimum resistance to the transfer of heat.

The 1989 revision of the provisions of Part L of Schedule 1 to The Building Regulations 1985 and Approved Document L set the maximum U value for the walls, ground floor and exposed floors of all buildings at 0.45 W/m²K and the maximum area of single glazed windows and roof lights at 15% of the total floor area for dwellings. For other building types the maximum single glazed areas are:

Building type	Windows (% exposed wall area)	Rooflights (% of roof area)
Other residential	25%	20%
Places of assembly, offices and shops	35%	20%
Industrial and storage	15%	20%

The stated purpose of the revision is to upgrade the requirements for the insulation of walls and roofs and by other revisions to offer greater flexibility in how the revised standards might be met.

In the calculation of heat loss an initial assumption is made that the aggregate area of windows and rooflights of dwellings is 15% of the total floor area and that the U value of the windows is 5.7, that is single glazing. Where the U value of the solid walls is less than the set limit then the area of the windows may be increased to the extent that the overall heat loss is not greater than it would be by accepting the initial assumption. Similarly if the U value of the windows is less than 5.7, by the use of double glazing, then the area of the windows can be increased or the

U value of the walls decreased, in what is termed a calculated 'trade off', providing the overall heat loss is not greater.

In imposing the need for higher standards of insulation it is recognised that there will be greater risks of condensation and thermal bridges (cold bridges) causing damage and inconvenience in buildings. The Building Research Establishment have, therefore, produced a document entitled *Thermal Insulation: Avoiding Risks* which will be referred to in the guidance notes in the Approved Document. To avoid the chance of stale air and of condensation due to warm, moist air associated with high levels of insulation and sealed windows the provisions of Part F of Schedule 1 of The Building Regulations, ventilation, have also been revised.

Because solid walls of brick or block offer poor resistance to the penetration of wind driven rain, the majority of load bearing walls are built with two leaves of brick or block separated by a cavity, whose prime purpose is as a barrier to rain penetration. The resistance to the passage of heat of a cavity of brick outer leaf, cavity and brick inner leaf with dense plaster finish is poor, having a U value of about 1.5 which is considerably greater than the 1989 revised maximum U value requirement of 0.45. To bring the wall up to the required resistance to the passage of heat it is necessary, therefore, to introduce a layer of some material with a high resistance to heat transfer, that is a low U value. Most of the materials that afford high resistance to heat transfer are fibrous or cellular, lightweight, have comparatively poor mechanical strength and are not suitable by themselves for use as either internal or external wall finishes. The logical position for such material in a cavity wall, therefore, is inside the cavity.

The purpose of the air space cavity in a cavity wall is as a barrier to the penetration of rainwater to the inside face of the wall. If the clear air space is to be effective as a barrier to rain penetration it should not be bridged by anythinig other than cavity ties. If the cavity is then filled with some insulating material, no matter how impermeable to water the material is, there will inevitably be narrow capillary paths around wall ties and between edges of boards or slabs across which water may penetrate. As a clear air space is necessary as a barrier to rain penetration there is good reason to fix insulation material inside a cavity so that it only partly fills the cavity and a cavity is maintained between the outer leaf and the insulating material. This construction, which is described as partial fill insulation of cavity, requires the use of some insulating material in the form of boards that are sufficiently rigid to be secured against the inner leaf of the cavity.

Cavity wall insulation

Partial fill

In theory a 25 wide air space between the outer leaf and the cavity insulation should be adequate to resist the penetration of rain providing the air space is clear of all mortar droppings and other building debris that might serve as a path for water. In practice, it is difficult to maintain a clear 25 wide air gap because of protrusion of mortar from joints in the outer leaf and the difficulty of keeping so narrow a space clear of mortar droppings. Good practice, therefore, is to use a 50 wide air space between the outer leaf and the partial fill insulation. With the decrease in maximum insulation requirements and the use of a 100 cavity with partial fill insulation it may be economic to use a lightweight block inner leaf to augment the insulation to bring the wall to the required U value.

Usual practice is to build the inner leaf of the cavity wall first, up to the first horizontal row of wall ties, then place the insulation boards in position against the inner leaf. Then as the outer leaf is built, a batten is suspended in the cavity air space and raised to the level of the first row of wall ties and the batten is then withdrawn and cleared of droppings. Insulation retaining wall ties are then bedded across the cavity to tie the leaves and retain the insulation in position and the sequence of operations is repeated at each level of wall ties.

The suspension of a batten in the air space and its withdrawal and cleaning at each level of ties does considerably slow the process of brick and block laying. If the cavity does serve a useful purpose then it is well worthwhile bearing the additional labour cost involved as a precaution against rain penetrating to the inner face of the wall.

Insulation retaining ties are usually standard galvanised steel or stainless steel wall ties to which a plastic disc is clipped to retain the edges of the insulation as illustrated in Fig. 104. Instead of building wall ties in, staggered one above the other, the ties may be set in line one over the other so that the retaining clips retain the corners of four insulation boards.

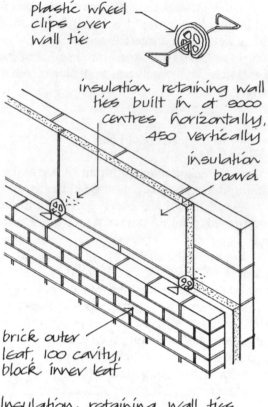

Insulation retaining wall ties to partial fill cavity wall

Fig. 104

The materials used for partial fill insulation should be of boards, slabs or batts that are sufficiently rigid for ease of handling and to be retained in a vertical position against the inner leaf inside the cavity without sagging or losing shape, so that the edges of the boards remain close butted throughout the useful life of the building. For small dwellings The Building Regulations do not limit the use of combustible materials as partial fill insulation in a cavity in a cavity wall.

To provide a clear air space of 50 inside the cavity as a barrier to rain penetration and to provide sufficient space to keep the cavity clear during building, an insulant with a low U value is of advantage if a nominal 75 wide cavity is formed between the outer and inner leaves.

Insulation materials

The materials used as insulation for the fabric of buildings may be grouped as inorganic and organic insulants.

Inorganic insulants are made from naturally occurring materials that are formed into fibre, powder or cellular structures that have a high void content, as for example, glass fibre, mineral fibre (rockwool), cellular glass beads, vermiculite, calcium silicate and magnesia or as compressed cork. Inorganic insulants are generally incombustible, do not support spread of flame, are rot and vermin proof and generally have a higher U value than organic insulants.

The inorganic insulants most used in the fabric of buildings are glass fibre and rockwool in the form of loose fibres, mats and rolls of felted fibres and semirigid and rigid boards, batts and slabs of compressed fibres, cellular glass beads fused together as rigid boards, compressed cork boards and vermiculite grains.

Organic insulants are based on hydrocarbon polymers in the form of thermosetting or thermoplastic resins to form structures with a high void content, as for example, polystyrene, polyurethane, isocyanurate and phenolic. Organic insulants generally have a lower U value than inorganic insulants, are combustible, support spread of flame more readily than inorganic insulants and have a comparatively low melting point.

The organic insulants most used for the fabric of buildings are expanded polystyrene in the form of beads or boards, extruded polystyrene in the form of boards and polyurethane, isocyanurate and phenolic foams in the form of preformed boards or spray coatings.

The materials that are cheapest, most readily available and used for cavity insulation are glass fibre, rockwool and EPS (expanded polytyrene), in the form of slabs or boards, in sizes to suit cavity tie spacing. With the recent increase in requirements for the insulation of walls it may well be advantageous to use one of the somewhat more expensive organic insulants such as XPS (extruded polystyrene), PIR (polyisocyanurate) or PUR (polyurethane), because of their lower U value, where a 50 clear air space is to be maintained in the cavity, without greatly increasing the overall width of the cavity.

Table 10 gives details of insulants made for use as partial fill to cavity walls.

Insulation thickness

A rough guide to determine the required thickness of insulation for a wall to achieve a U value of 0.45 is to

Table 10. Insulating Materials

Cavity wall partial fill	Thickness	U valve W/m²K
Glass fibre rigid slab 455 × 1200	30, 35, 40, 45	0.033
Rockwool rigid slab 455 × 1200	30, 40, 50	0.033
Cellular glass 450 × 600	40, 45, 50	0.042
EPS boards 450 × 1200	25, 40, 50	0.037
XPS boards 450 × 1200	25, 30, 50	0.028
PIR boards 450 × 1200	20, 25, 30, 35, 50	0.022
PUR boards 450 × 1200	20, 25, 30, 35, 40, 50	0.022

EPS expanded polystyrene
XPS extruded polystyrene
PIR rigid polyisocyanurate
PUR rigid polyurethane

assume the insulant provides the whole or a major part of the insulation by using 30 thickness with a U value of 0.02, 46 with 0.03, 61 with 0.04, 76 with 0.05 and 92 with 0.06.

A more exact method is by calculation as follows.

$$\text{Thermal resistance required} = \frac{1}{0.45} = 2.22 \text{ m}^2\text{K/W}$$

Thermal resistance of construction is:

External surface	0.06 m²K/W
102 brick outer leaf	0.12
cavity at least 50	0.18
115 block inner leaf	1.05
13 plasterboard	0.03
inside surface	0.12
Total	1.56

Additional resistance to be provided by insulation

$$= 2.22 - 1.56$$
$$= 0.66 \text{ m}^2\text{K/W}$$

Assume insulation with a U value of 0.03
Thickness of insulation required $= 0.66 \times 0.03 \times 1000$
$$= 19.8 \text{ mm}$$

This thickness is about one-third of that proposed by the first method that takes little account of the resistance of the rest of the wall construction.

Total fill

The thermal insulation of external walls by totally filling the cavity has been in use for many years. There have been remarkably few reported incidents of penetration of water through the total fill of cavities to the inside face of walls and the system of total fill has become an accepted method of insulating cavity walls.

The method of totally filling cavities with an insulant was developed after the steep increase in the price of oil and other fuels in the mid-1960s, as being the most practical way to improve the thermal insulation of existing cavity walls. Small particles of glass or rock wool fibre or foaming organic materials were blown through holes drilled in the outer leaf of existing walls to completely fill the cavity. This system of totally filling the cavity of existing walls has been very extensively and successfully used. The few reported failures due to penetration of rainwater to the inside face, were due to poor workmanship in the construction of the walls. Water penetrated across wall ties sloping down into the inside face of the wall, across mortar droppings bridging the cavity or from mortar protruding into the cavity from the outer leaf.

From the few failures due to rain penetration it would seem likely that the cavity in existing walls that have been totally filled, was of little, if any, critical importance in resisting rain penetration in the position of exposure in which the walls were situated. None the less it is wise to provide a clear air space in a cavity wherever practical, against the possibility of rain penetration.

Where insulation is used to totally fill a nominal 50 wide cavity there is no need to use insulation retaining wall ties.

With a brick outer and block inner leaf it is preferable to raise the outer brick leaf first so that mortar protrusions from the joints, sometimes called snots, can be cleaned off before the insulation is placed in position and the inner block leaf, with its more widely spaced joints is built, to minimise the number of mortar snots that may stick into the cavity. This sequence of operations will require scaffolding on both sides of the wall and so add to the cost.

Insulation that is built in as the cavity walls are raised, to totally fill the cavity, will to an extent be held in position by the wall ties and the two leaves of the cavity wall. Rolls or mats of loosely felted glass fibre or rockwool are often used. There is some little likelihood that these materials may sink inside the cavity and gaps may open up in the insulation and so form cold bridges across the wall. To maintain a continuous, vertical layer of insulation inside the cavity one of the mineral fibre semi-rigid batts or slabs should be used. Fibre glass and rockwool semi-rigid batts or slabs in sizes suited to cavity tie spacing are made specifically for this purpose. As the materials are made in widths to suit vertical wall tie spacing there is no need to push them down into the cavity after the wall is built, as is often the procedure with loose fibre rolls and mats, and so displace freshly laid brick or blockwork. There is no advantage in using one of the more expensive organic insulants such as XPS, PIR or PUR that have a lower U value than mineral fibre materials for the total cavity fill, as the width of the cavity can be adjusted to suit the required thickness of insulation.

The most effective way of insulating an existing cavity wall is to fill the cavity with some insulating material that can be blown into the cavity through small holes drilled in the outer leaf of the wall. The injection of the cavity fill is a comparatively simple job. The complication arises in forming sleeves around air vents penetrating the wall and sealing gaps around openings.

When filling the cavity of existing walls became common practice, a foamed organic insulant, urea-formaldehyde, was extensively used. The advantage of this material was that it could be blown, under pressure, through small holes in the outer leaf and as the constituents mixed they foamed and filled the cavity with an effective insulant. This material was extensively used, often by operatives ill-trained in the sensible use of the material. The consequence was that through careless mixing of the components of the insulant and careless workmanship, the material gave off irritant fumes when used and later when it was in place, these entered buildings and caused considerable distress to the occupants. Approved Document D to Schedule 1 of The Building Regulations 1985, details provisions for the use of this material in relation to the construction of the wall and its suitability, the composition of the materials, and control of those carrying out the work. As a result of past failures this material is less used than it was.

Glass fibre, granulated rockwool or EPS beads are used for the injection of insulation of existing cavity walls. These materials can also be used for blowing into the cavity of newly built walls.

Table 11 gives details of insulants for total cavity fill.

The required thickness of insulation can be taken from the two methods suggested in the 1989 revision of The Building Regulations described on page 78 for partial fill in relation to the U value of the chosen material or similarly by calculation. In a calculation for total fill, the thermal resistance of the cavity is omitted.

Table 11. Insulating Materials

Cavity wall total fill	Thickness	U valve W/m²K
Glass fibre semi-rigid batt 455 × 1200	50, 65, 75, 100	0.036
glass fibres for blown fill		0.039
Rockwool semi-rigid batt 455 × 900	50, 65, 75, 100	0.036
granulated for blown fill		0.037
EPS beads for blown fill		0.04

EPS expanded polystyrene

Thermal bridge: A thermal bridge, more commonly known as a cold bridge in cold climates, is caused by appreciably greater thermal conductivity through one part of a wall than the rest of the wall. Where the cavity in a cavity wall is partially or totally filled with insulation and the cavity is bridged with solid filling at the head, jambs or cill of an opening, there will considerably greater transfer of heat through the solid filling than through the rest of the wall. Because of the greater transfer of heat through the solid filling illustrated in Fig. 105, the inside face of the wall will be appreciably colder in winter, than the rest of the wall and cause some loss of heat and encourage warm moist air to condense on the inside face of the wall on the inside of the cold bridge. This condensation water may cause unsightly stains around openings and encourage mould growth.

Thermal bridges can be minimised by continuing cavity insulation down to the head of windows and doors and up to the sides and bottom of doors and windows as illustrated in Fig. 105.

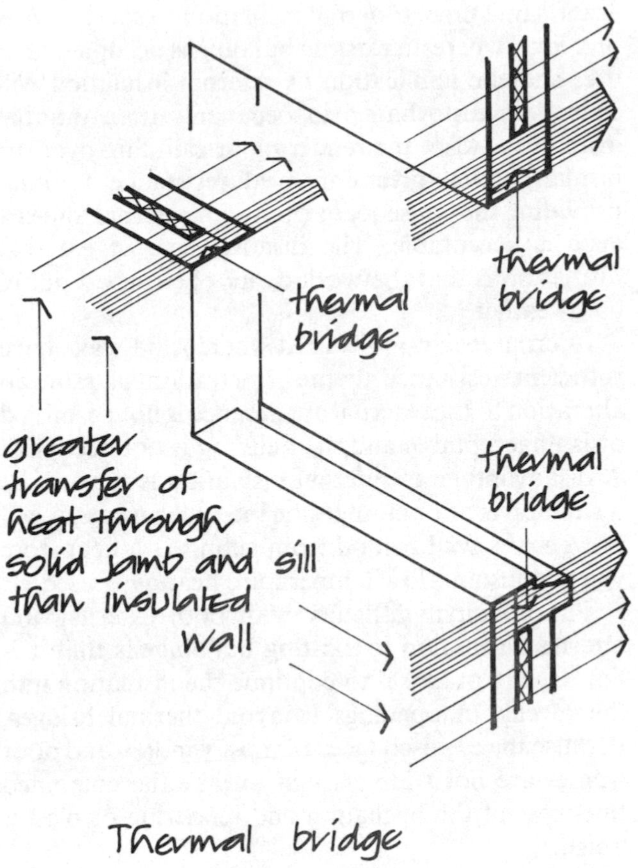

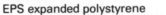

greater transfer of heat through solid jamb and sill than insulated wall

Thermal bridge

Fig. 105

Solid walls of brick and block

Solid walls of brick and block offer poor resistance to the passage of heat. The U value, that is the thermal transmittance coefficient, of a brick wall 220 thick with a 13 finish of dense plaster is 2.1, which is considerably in excess of the maximum U value of 0.45 for walls. To reduce the heat loss through solid walls of brick or block it is necessary to fix a material with a low U value either to the inside or the outside of the wall. The choice of internal or external insulation to a solid wall depends on the resistance of the wall to rain penetration, its external appearance and whether the building is occupied or unoccupied.

The resistance of a solid wall to rain penetration is greatly improved by an external facing of tile, slate, sheet metal or plastic cladding and to a lesser extent by rendering. Where external cladding or rendering is to be used, or is proposed, to enhance resistance to rain penetration it is advantageous to incorporate a lining of insulation behind the cladding or rendering as external insulation to improve thermal insulation at the same time. External insulation to solid walls is preferred where an existing building is occupied or in use, and the application of external insulation will cause less disturbance to occupants than internal insulation. Also, the rendering or cladding over the insulation will give improved resistance to rain, providing the consequent change in external appearance is acceptable. The disadvantage of external insulation is that the work cannot be carried out in bad weather.

Internal insulation is used where solid walls have sufficient resistance to the penetration of rain, an alteration to the external appearance is not permitted or is unacceptable and the building is not occupied. A disadvantage of internal insulation is that as the insulation is at or close to the internal surface, it will prevent the wall behind from acting as a heat store where constant, low temperature heating is used.

The principal difficulty with both external and internal insulation to existing buildings is that it is not usually practical to continue the insulation into the reveals of openings to avoid thermal bridges, because the exposed faces of most window and door frames are not wide enough to take the combined thickness of the insulation and rendering or plaster finish.

Newly constructed walls can be formed with outward facing rebates to openings into which the insulation can be turned or subframes formed for windows and doors for the same purpose.

External insulation

Insulating materials by themselves do not provide a satisfactory external finish to walls against rain penetration or for appearance sake and have to be covered with a finish of cement rendering, paint or a cladding material such as tile, slate or weatherboarding. For rendered finishes, one of the inorganic insulants rockwool or cellular glass in the form of rigid boards are most suited. For cladding, one of the organic insulants such as XPS, PIR or PUR are used for their low U values that necessitates least thickness of board.

As a base for applied rendering the insulation boards or slabs are first bedded and fixed in line on dabs of either gap filling organic adhesive or dabs of polymer emulsion mortar and secured with corrosion resistant fixings to the wall. As a key for the render coats, either the insulation boards have a keyed surface or expanded metal lath or glass fibre

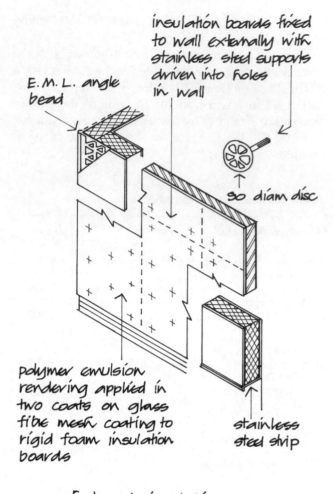

insulation boards fixed to wall externally with stainless steel supports driven into holes in wall

E.M.L. angle bead

90 diam disc

polymer emulsion rendering applied in two coats on glass fibre mesh coating to rigid foam insulation boards

stainless steel strip

External insulation

Fig. 106

mesh is applied to the face of the insulation. The weather protective render is applied in two coats by traditional wet render application, by rough casting or by spray application and finished smooth, coarse or textured. Coarse, spatter dash or textured finishes are preferred as they disguise hair cracks that are due to drying shrinkage of the rendering.

Because the rendering is applied over a layer of insulation it will be subject to greater temperature fluctuations than it would be if applied directly to a wall, and so is more liable to crack. To minimise cracking due to temperature change and moisture movements, the rendering should be reinforced with a mesh securely fixed to the wall, and movement joints should be formed at not more than 6 m intervals. The use of a light coloured finish and rendering incorporating a polymer emulsion, will reduce cracking.

As the overall thickness of the external insulation and rendering is too great to be returned into the reveals of existing openings it is usual to return the rendering by itself, or fix some non-ferrous or plastic trim to mask the edge of the insulation and rendering. The reveals of openings will act as thermal bridges to make the inside face of the wall around openings colder than the rest of the wall. Fig. 106 is an illustration of insulated rendering applied externally.

Tile and slate hanging, timber weatherboarding and profiled sheets can be fixed over a layer of insulating material behind the battens or sheeting rails to which these cladding materials are fixed.

Slabs of compressed rockwool are cut and shaped with bevel edges to simulate the appearance of masonry blocks. The blocks are secured to the external face of the wall with stainless steel brackets, fixed to the wall to support and restrain the blocks that are arranged with either horizontal, bonded joints or vertical and horizontal continuous joints. An exterior quality paint is then applied to the impregnated surface of the blocks. At openings, non-ferrous or plastic trim is fixed around outer reveals.

Details of insulating materials are given in table 12.

Table 12. Insulating Materials

Solid wall external insulation	Thickness	*U* valve W/m²K
Rockwool rigid slab with polymer cement finish	30, 40, 50, 60, 75, 100	0.033
Cellular glass boards for rendering on EML	30, 35, 40, 50, 60, 70	0.042
XPS boards T & G on long edges for cladding	25, 50	0.025
PIR boards behind cladding	25, 30, 35, 40	0.022
PUR boards behind cladding	20, 25, 30, 35, 40, 50	0.022

XPS extruded polystyrene
PIR rigid polyisocyanurate
PUR rigid polyurethane

Internal insulation

Internal insulation can be used to improve the thermal resistance of both new and old solid and cavity walls. Its principal use is as an inner insulating lining to existing solid walls. Insulating materials are lightweight and do not generally have a smooth hard finish and are not, therefore, suitable as the inside face of the walls of most buildings. It is usual to cover the insulating layer with a lining of plasterboard or plaster so that the combined thickness of the inner lining and the wall have a *U* value of 0.45 or less.

Internal linings for thermal insulation are either of preformed, laminated panels that combine a wall lining of plasterboard glued to an insulation board with a moisture vapour barrier on the wall side or of separate insulation material that is fixed to the wall and then covered with plasterboard or wet plaster. The method of fixing the lining to the inside wall surface depends on the surface to which it is applied.

Fixing

Adhesive fixing directly to the inside wall face is used for preformed, laminated panels and for rigid insulation boards. Where the inside face of the wall is clean, dry, level and reasonably smooth, as for example, a sound plaster finish or a smooth and level concrete, brick or block face, the laminated panels or rigid insulation boards are secured with organic based, gap filling adhesive that is applied in dabs and strips to the back of the boards or panels or to both the boards and wall. The panels or boards are then applied and pressed into position against the wall face and their position adjusted with a foot lifter. Where the surface of the wall to be lined is uneven or rough the laminated panels or insulation boards are fixed with dabs of plaster bonding, applied to both the wall surface and the back of the lining. Dabs are small areas of wet plaster bonding applied at intervals on the surface with a trowel, as a bedding and adhesive. The lining is applied and pressed into position against the wall. The wet dabs of bonding allow for irregularities in the wall surface and also serve as an adhesive. Some of the lining systems use secondary fixings in addition to adhesive. These secondary fixings are non-ferrous or plastic nails or screws driven or screwed through the insulation boards into the wall.

Mechanical fixing: As an alternative to adhesive fixing, the insulating lining and the wall finish can be fixed to wood battens that are nailed to the wall with packing pieces as necessary, to form a level surface. The battens should be impregnated against rot and fixed with non-ferrous fixings. The insulating lining is fixed either between the battens or across the battens and an internal lining of plasterboard is then nailed to the battens, through the insulation.

The thermal resistance of wood is less than that of most insulating materials. When the insulating material is fixed between the battens there will be cold bridges through the battens that may cause pattern staining on wall faces.

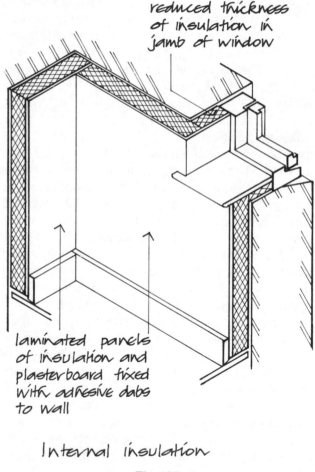

reduced thickness
of insulation in
jamb of window

laminated panels
of insulation and
plasterboard fixed
with adhesive dabs
to wall

Internal insulation

Fig. 107

Internal finish: An inner lining of plasterboard can be finished by taping and filling the joints or with a thin skim coat of neat plaster. A plaster finish of lightweight plaster and finishing coat is applied to the ready keyed surface of some insulating boards or to

expanded metal lathing fixed to battens. (For details of plaster, see Volume 2.)

Laminated panels of insulation lined on one side with a plasterboard finish are made specifically for the insulation of internal walls. The panels are fixed with adhesive or mechanical fixings to the inside face of the wall. For internal lining the organic insulants such as XPS, PIR and PUR have the advantage of least thickness of material necessary due to their low U value.

Fig. 107 is an illustration of preformed, laminated, insulating, lining panels fixed to the inside of a solid wall.

Table 13 gives details of materials for internal insulation.

Table 13. Insulating Materials

Solid wall internal insulation	Thickness	U valve W/m²K
Glass fibre laminated panel glass fibre slab and plasterboard	30, 40, 60	0.031
Rockwool laminated panel rockwool slab and plasterboard	25, 32, 40, 50	0.033
EPS laminated panel EPS board and plasterboard	25, 32, 40, 50	0.037
XPS boards keyed for plaster	25, 50, 75, 110	0.033
PIR boards reinforced with glass fibre tissue both sides	25, 30, 35, 40	0.022
PUR laminated panel PUR board and plasterboard	12.5, 15, 20, 25, 30	0.022

EPS expanded polystyrene
XPS extruded polystyrene
PIR rigid polyisocyanurate
PUR rigid polyurethane

Vapour barriers, vapour check: The moisture vapour pressure from warm moist air inside insulated buildings may find its way through internal linings and condense to water on the cold outer face of the insulation material. Where the condensation moisture is absorbed by the insulation it will reduce the efficiency of the insulation and where condensation saturates battens, they may rot. With insulation that is permeable to moisture vapour, a vapour barrier or check should be fixed on the room side of insulation. A vapour barrier is one that completely stops the movement of vapour through it and a vapour check is one that substantially stops vapour. As it is difficult to make a complete seal across the whole surface of a wall including all overlaps of the barrier and at angles, it is in effect impossible to form a barrier and the term vapour check should more properly be used. Sheets of polythene with edges overlapped are commonly used as a vapour check. Some insulating materials are impermeable to moisture vapour and do not need a vapour check, providing the edges of panels or boards of these materials can be tightly butted together.

Resistance to airborne and impact sound

The requirement in Part E of Schedule 1 to The Building Regulations 1985 is that walls which separate a dwelling from another building or from another dwelling shall have reasonable resistance to airborne sound.

Where solid walls of brick or block are used to separate dwellings the reduction of airborne sound between dwellings depends mainly on the weight of the wall and its thickness. A cavity wall with two leaves of brick or block does not afford the same sound reduction as a solid wall of the same equivalent thickness because the stiffness of the two separate leaves is less than that of the solid wall and in consequence is more readily set into vibration. The joints between bricks or blocks should be solidly filled with mortar and joints between the top of a wall and ceilings should be filled against airborne sound transmission.

In Approved Document E, giving practical guidance to meeting the requirements of The Building Regulations 1985 in relation to walls between dwellings, is a table giving the minimum weight of walls to provide adequate airborne sound reduction. For example, a solid brick wall 215 thick, plastered both sides, should weigh at least 300 kg/m^2 including plaster, and a similar cavity wall 255 thick, plastered both sides, should weigh at least 415 kg/m^2 including plaster. A solid concrete block wall 215 thick, plastered both sides should weigh at least 415 kg/m^2 including plaster, and a cavity block wall 250 thick, plastered both sides, should weigh at least 415 kg/m^2 including plaster.

STONE MASONRY WALLS

Before the industrial revolution, many permanent buildings in hill and mountain districts and most large buildings in lowland areas in this country were built of natural stone. At that time the supply of stone from local quarries was adequate for the buildings of the small population of this country. The increase in population that followed the industrial revolution was so great that the supply of sound stone was quite inadequate for the new buildings being put up. Coal was cheap, the railway spread throughout the country and cheap mass produced bricks largely replaced stone as the principal material for the walls of all but larger buildings.

Because natural stone is expensive it is principally used today as a facing material bonded or fixed to a backing of brickwork or concrete (see Volume 4). Many of the larger civic and commercial buildings are faced with natural stone because of its durability, texture, colour and sense of permanence. Natural stone is also used as the outer leaf of cavity walls for houses in areas where local quarries can supply stone at reasonable cost.

In recent years much of the time consuming and therefore expensive labour of cutting, shaping and finishing building stone has been appreciably reduced by the use of power operated tools, edged or surfaced with diamonds. This facility has improved output in the continuing and extensive work of repair and maintenance to stone buildings and encouraged the use of natural stone as a facing material.

Because natural stone is an expensive material, cast stone has been used as a cheaper substitute. Cast stone is made from either crushed natural stone or natural aggregate and cement and water which is cast in moulds. The cast stone blocks are made to resemble natural stone. When used in walling, cast stone often cracks and becomes dirt stained in an irregular, unsightly way.

Natural stone

The natural stones used in building may be classified by reference to their origin as:

(a) Igneous
(b) Sedimentary
(c) Metamorphic

Igneous stones were formed by the cooling of molten magma as the earth's crust cooled, shrank and folded to form beds of igneous rock. Of the igneous stones that can be used for building such as granite, basalt, diorite and serpentine, granite is most used for walls of buildings.

Granite consists principally of crystals of felspar which is made up of lime and soda with other minerals in varying proportions and small grains of quartz and mica which give a sparkle to the surface of the stone. The granite that is native to these islands that is most used for walling is sometimes loosely described as Aberdeen granite as it is mined from deep beds of igneous rock near that town in Scotland. The best known Aberdeen granites are Rubislaw which is blue grey, Kemnay which is grey and Peterhead which is pink in colouring. All of these granites are fine grained, hard and durable and can be finished to a smooth polished surface. Aberdeen granites have been much used for their strength and durability as a walling material for large buildings and are now used as a facing material.

Cornish and Devon granites are coarse grained, light grey in colour with pronounced grains of white and black crystals visible. The stone is very hard and practically indestructible. Because these granites are coarse grained and hard they are laborious to cut and shape and cannot easily be finished with a fine smooth face. These granites have been principally used in engineering works for bridges, lighthouses and docks and also as a walling material for buildings in the counties of their origin.

Sedimentary stone

Sedimentary stone was formed gradually over thousands of years from the disintegration of older rocks which were broken down by weathering and erosion or from accumulations of organic origin, the resulting fine particles being deposited in water in which they settled in layers, or by being spread by wind in layers that eventually consolidated and hardened to form layers of sedimentary rocks and clays. Because sedimentary stone is formed in layers it is said to be stratified. The strata or layers make this type of stone easier to split and cut then hard igneous stones that are not stratified. The strata also affect the way in which the stone is used, if it is to be durable, as the divisions between the layers or strata are, in effect, planes of weakness.

A general subdivision of sedimentary stones is as:

Limestone

Sandstone

The limestones used for walling consist mainly of grains of shell or sand surrounded by calcium carbonate, which are cemented together with calcium carbonate. The limestones most used for walling are quarried from beds of stone in the southwest of England, those most used being Portland and Bath stone. Because limestone is a stratified rock, due to the deposit of layers, it must be laid on its natural bed in walls.

Portland stone is quarried in Portland Island on the coast of Dorsetshire. There are extensive beds of this stone which is creamy white in colour, weathers well and used to be particularly popular for walling for larger buildings in towns. Many large buildings have been built of Portland stone because an adequate supply of large stones is available, the stone is fine grained and delicate mouldings can be cut on it and it weathers well even in industrial atmospheres. Among the buildings constructed with this stone are the great Banqueting Hall in Whitehall (1639), St Paul's Cathedral (1676), the British Museum (1753) and Somerset House (1776). More recently, many large buildings have been faced with this stone.

In the Portland stone quarries are three distinct beds of the stone, the base bed, the whit bed and the roach. The base bed is a fine, even grained stone which is used for both external and internal work to be finished with delicate mouldings and enrichment.

The whit bed is a hard, fairly fine grained stone which weathers particularly well, even in towns whose atmosphere is heavily polluted with soot and it is extensively used as a facing material for large buildings.

The roach is a tough, coarse grained stone which has principally been used for marine construction such as piers and lighthouses.

The stones from the different beds of Portland limestone look alike to the layman. It is sometimes difficult for even the trained stonemason to distinguish base bed from whit bed. Roach can be distinguished by its coarse grain and by the remains of fossil shells embedded in it. When taken from the quarry the stone is moist and comparatively soft, but gradually hardens as moisture (quarry sap) dries out.

Bath stone: Many of the building in the town of Bath were built with a limestone quarried around the town. This limestone is one of the great Oolites and a similar stone was also quarried in Oxfordshire. Bath stone from the Taynton (Oxfordshire) quarry was extensively used in the construction of the early colleges in Oxford (St Johns, for example) during the twelfth, thirteenth and fourteenth centuries. Many of the permanent buildings in Wiltshire and Oxfordshire were built of this stone which varies from fine grained to coarse grained in texture and light cream to buff in colour. Most of the original quarries are no longer being worked.

The durability of Bath stone varies considerably. Some early buildings constructed with this stone are well preserved to this day but others have so decayed over the years and been so extensively repaired that little of the original stone remains. Extensive repair of the Bath stone fabric of several of the colleges in Oxford has been carried out and continuing repair is necessary.

Sandstone was formed from particles of rock broken down over thousands of years by the action of winds and rain. The particles were washed into and settled to the beds of lakes and seas in combination with clay, lime and magnesia and gradually compressed into strata of sandstone rock. The particles of sandstone are practically indestructible and the hardness and resistance to the weather of this stone depends on the composition of the minerals binding the particles of sand. If the sand particles are bound with lime the stone often does not weather well as the soluble lime dissolves and the stone disintegrates. The material binding the sand particles should be insoluble and crystalline. Sandstones are generally coarse grained and cannot be worked to fine mouldings.

The stratification of most sandstones is visible as fairly close spaced divisions in the sandy mass of the stone. It is essential that this type of stone be laid on its natural bed in walls.

Most sandstones have been quarried in the northern counties of England where for centuries this stone has been the material commonly used for the walls of buildings. Some of the sandstones that have been commonly used are:

Crosland Hill (Yorkshire). A light brown sandstone of great strength which weathers well and is used for masonry walls as a facing material and for engineering works. It is one of the stones known as hard York stone, a general term used to embrace any hard sandstone not necessarily quarried in Yorkshire.

Blaxter stone (Northumberland). A hard, cream coloured stone used for walling and as a facing.

Doddington (Northumberland). A hard, pink stone used for walling.

Darley Dale (Derbyshire). A hard, durable stone of great strength much used for engineering works and as walling. It is hard to work and generally used in plain, unornamented walls. Buff and white varieties of this stone were quarried.

Forest of Dean (Gloucestershire). A hard, durable, grey or blue grey stone which is hard to work but weathers well as masonry walling.

Metamorphic stones

Metamorphic stones were formed from older stones that were changed by pressure or heat or both. The metamorphic stones used in building are slate and marble.

Slate was formed by immense pressure on beds of clay that were compressed to hard, stratified slate which is used for roofing and as sills and copings in building. Riven, split, Welsh slate has for centuries been one of the traditional roofing materials used in this country. The stone can be split to comparatively thin slates that are hard and very durable.

Marble: The description marble is used to include many stones that are not true metamorphic rocks, such as limestones that can take a fine polish. In the British Isles true marble is only found in Ireland and Scotland. Marble is principally used as an internal facing material in this country.

Durability of natural stone

Natural stone has been used in the construction of buildings because it was thought that any hard, natural stone would resist the action of wind and rain for centuries. Many natural stones have been used in walling and have been durable for a hundred or more years and are likely to have a comparable life if reasonably maintained. There have been some notable failures of natural stone in walling, due in the main to a poor selection of the material and poor workmanship. The best known example of decay in stonework occurred in the fabric of the Houses of Parliament, the walls of which were built with a magnesian limestone from Ancaster in Yorkshire. A Royal Commission reported in 1839 that the magnesian limestone quarried at Bolsover Moor in Yorkshire was considered the most durable stone to be obtained for the Houses of Parliament. After building work had begun it was discovered that the quarry was unable to supply sufficient large stones for the building and a similar stone from the neighbouring quarry at Anston was chosen as a substitute. The quarrying, cutting and use of the stone was not supervised closely and in consequence many inferior stones found their way into the building and many otherwise sound stones were incorrectly laid. Decay of the fabric has been continuous since the Houses of Parliament were first completed, and extensive, costly renewal of stone has been going on for many years. At about the same time that the Houses of Parliament were being built, the Museum of Practical Geology was built in London of Anston stone from the same quarry that supplied the stone for the Houses of Parliament, but the quarrying, cutting and use of the stones was closely supervised for the museum, whose fabric remained sound. The variability of natural stone that may appear sound and durable but some of which may not weather well is one of the disadvantages of this material which can, when carefully selected and used, be immensely durable and attractive as a walling material.

Seasoning natural stone: Some natural stones are comparatively soft and moist when first quarried but gradually harden. Building stones should be seasoned (allowed to harden) for periods of up to a few years, depending on the size of the stones. Once stone has been seasoned it does not revert to its original soft moist state on exposure to rain, but on the contrary hardens with age.

Bedding stones: Natural stones that are stratified, limestone and sandstone, must be used in walling so that they lie on their natural bed to support compressive stress. The bed of a stone is its face parallel to the strata (layer) of the stones in the quarry and the stress that the stone suffers in use should be at right angles to the strata or bed which otherwise might act as a plane of weakness and give way under compressive stress. The stones in an arch are laid with the bed or strata radiating roughly from the centre of the arch so that the bed is at right angles to the compressive stress acting around the curve of the arch.

Cast stone

Cast stone is the term used to describe concrete cast in moulds to resemble blocks of natural stone. The concrete mix used may be an aggregate of crushed natural stone or an aggregate of gravel and sand, which is mixed with cement and water.

Reconstructed stone is made from an aggregate of crushed stone, cement and water. The stone is crushed so that the maximum size of the particles is 15 and it is mixed with cement in the proportions of 1 part of cement to 3 or 4 parts of stone. Either Portland cement, white cement or coloured cement may be used to simulate the colour of a natural stone as closely as possible. A comparatively dry mix of cement, crushed stone and water is prepared and cast in wood moulds. The mix is thoroughly consolidated inside the moulds by vibrating and left to harden in the moulds for at least 24 hours. The stones are then taken out of the moulds and allowed to harden gradually for twenty-eight days.

Well-made reconstructed stone has much the same texture and colour as the natural stone from which it is made and can be cut, carved and dressed just like natural stone. It is not stratified, is free from flaws and is sometimes a better material than the natural stone from which it is made. The cost of a plain stone, cast with an aggregate of crushed natural stone is about the same as that of a similar natural stone. Moulded cast stones can often be produced more cheaply by repetitive casting than similar natural stones that have to be cut and shaped by hand.

A cheaper form of cast stone is made with a core of ordinary concrete, faced with an aggregate of crushed natural stone and cement. The core is made from clean gravel, sand and Portland cement and the facing from crushed stone and cement to resemble the texture and colour of a natural stone. The crushed stone, cement and water is first spread in the base of the mould to a thickness of about 25, the core concrete is added and the mix consolidated. If the stone is to be exposed on two or more faces the natural stone mix is spread up the sides and the bottom of the mould. This type of cast stone obviously cannot be carved as it has only a thin surface of natural looking stone.

As an alternative to a facing of reconstructed stone, the facing or facings can be made of cement and sand pigmented to look somewhat like the colour of a natural stone.

Cast stone that is made from too wet a mix and stones that are not allowed to harden gradually, are likely to suffer surface cracking due to the shrinkage of the cement rich material. This surface cracking or crazing is unsightly, encourages irregular dirt staining and frost action may in time cause the surface to spall and disintegrate.

Cast stone is used as an ashlar facing to brick or concrete backgrounds and as blocks for the outer leaf of cavity walls as a substitute for natural stone. Cast stone does not weather in the same way that sedimentary stone does by a gradual change of colour. The material tends to have a lifeless, mechanical appearance and will in a short time tend to show irregular, unsightly dirt stains at joints, cracks and around projections. It is by and large, a very poor substitute for the natural material.

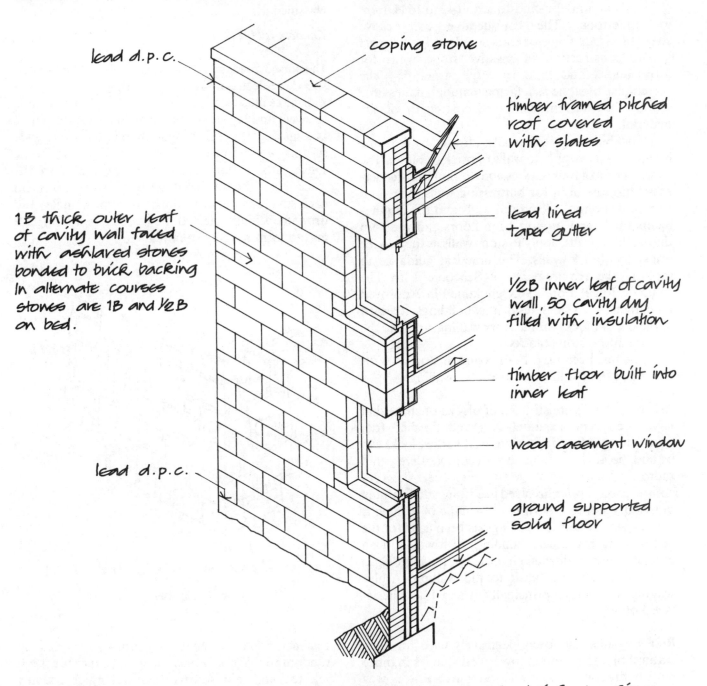

lead d.p.c.

coping stone

timber framed pitched roof covered with slates

lead lined taper gutter

1B thick outer leaf of cavity wall faced with ashlared stones bonded to brick backing In alternate courses stones are 1B and ½B on bed.

½B inner leaf of cavity wall, 50 cavity dry filled with insulation

timber floor built into inner leaf

wood casement window

lead d.p.c.

ground supported solid floor

Cavity wall faced with ashlared stone bonded to brick backing

Fig. 108

Functional requirements

Strength and stability: The strength of sound building stone lies in its very considerable compressive strength. The ultimate or failing stress of stone used for walling is about 300 to 100 N/mm^3 for granite, 195 to 27 N/mm^3 for sandstone and 42 to 16 N/mm^3 for limestone. The considerable compressive strength of building stone was employed in the past in the construction of massive stone walls for fortifications and in other large structures. The current use of stone as a facing material makes little use of the inherent compressive strength of the material.

The stability of a stone wall is affected by the same limitations that apply to walls of brick or block. The construction of foundations and the limits of slenderness ratio, the need for buttressing walls, piers and chimneys along the length of walls and the requirements for lateral support from floors and roofs up the height of walls, apply to stone walls as they do for brick and block walls. The practical guidance to meeting the requirements of Schedule 1 to The Building Regulations 1985, contained in Approved Document A for small residential buildings in regard to the use of blocks for masonry walling, apply to the use of natural stone blocks.

Stone has been used in the construction of walls either as ashlar walling or as rubble walling.

Ashlar walling is constructed of blocks of stone that have been very accurately cut and finished true square to specified dimensions so that the blocks can be laid, bedded and bonded with comparatively thin mortar joints as illustrated in Fig. 108. The very considerable labour involved in cutting and finishing individual stones is such that this type of walling is very expensive. Ashlar walling has been used for the larger more permanent buildings in towns and on estates where the formal character of the building is pronounced by the finish to the walling. Ashlar walling is now used principally as a facing material (see Volume 4).

Rubble walling has been extensively used for agricultural buildings and in towns and villages in those parts of the country where a local source of stone was readily available. The term rubble describes blocks of stone as they come from the quarry. The rough rubble stones are used in walling with little cutting other than the removal of inconvenient corners. The

various types of rubble walling depend on the nature of the stone used. Those stones that are hard and laborious to cut or shape are used as random rubble and those sedimentary stones that come from the quarry roughly square are used as squared rubble.

The various forms of rubble walling may be classified as:

Random rubble

Uncoursed random rubble stones of all shapes and sizes are selected more or less at random and laid in mortar as illustrated in Fig. 109. No attempt is made to select and lay stones in horizontal courses. There is some degree of selection to avoid excessively wide mortar joints and also to bond stones by laying some longer stones both along the face and into the thickness of the wall, so that there is a bond stone in each square metre of walling. At quoins, angles and around openings selected stones or shaped stones are laid to form roughly square angles.

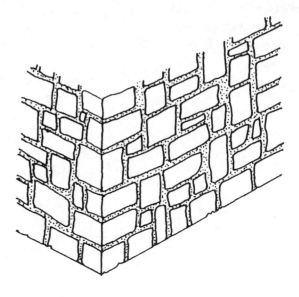

Random rubble uncoursed

Fig. 109

Random rubble brought to courses is similar to random rubble uncoursed except that the stones are selected and laid so that the walling is roughly levelled in horizontal courses at vertical intervals of from 600 to 900 as illustrated in Fig. 110. As with uncoursed rubble transverse and longitudinal bond stones are used.

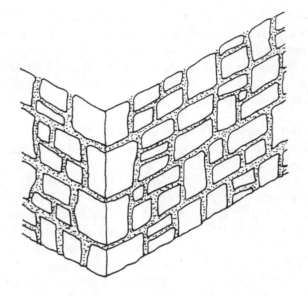

Random rubble coursed

Fig. 110

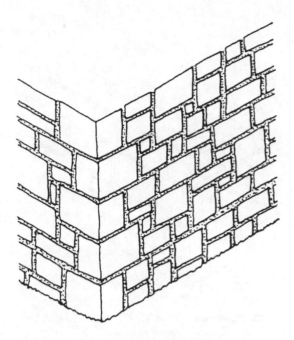

Squared rubble uncoursed

Fig. 111

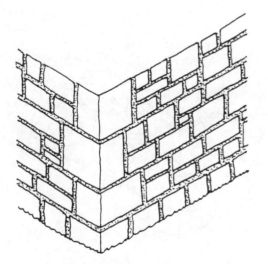

Squared rubble brought to courses

Fig. 112

Squared rubble

Squared random rubble uncoursed is laid with stones that come roughly square from the quarry in a variety of sizes. The stones are selected at random, are roughly squared with a walling hammer and laid without courses as illustrated in Fig. 111. As with random rubble both transverse and longitudinal bond stones are laid at intervals.

Snecked rubble is a term for squared random rubble in which a number of small squared stones, snecks, are laid to break up long continuous vertical joints. Snecked rubble is often difficult to distinguish from squared random rubble.

Squared rubble brought to courses is constructed from roughly square stone rubble, selected and squared so that the work is brought to courses every 300 to 900 intervals as illustrated in Fig. 112.

Squared rubble coursed is built with stones that are roughly squared so that the stones in each course are roughly the same height and the courses vary in height as illustrated in Fig. 113. The face of the stones may be roughly dressed to give a rockfaced appearance or dressed smooth to give a more formal appearance.

91

Polygonal walling or 'Kentish Rag': Stones that are taken from a quarry where the stone is hard, has no pronounced laminations and comes in irregular shapes can be laid as polygonal walling. The stones are selected and roughly dressed to fit when laid, to an irregular pattern as illustrated in Fig. 114.

Flint walling is traditional to East Anglia and the south and south-east of England. Both field and shore flints are used. The flints used for walling are up to 300 in length and from 75 to 150 in width and thickness. The flints or cobbles may be used whole or split to show the heart of the flint and also knapped or snapped so that they show a roughly square face. Flint walling is built with a dressing of stone or brick at angles and in horizontal lacing courses that level the wall at intervals. Figure 115 is an illustration of whole flints laid without courses in brick dressing to angles and as lacing and Fig. 116, an illustration of knapped flints laid to courses in stone dressing.

Rubble walling is generally considerably thicker than a wall of square bricks or blocks because of the inherent instability of a wall built of irregular shaped blocks. Rubble walling for buildings is usually at least 400 thick.

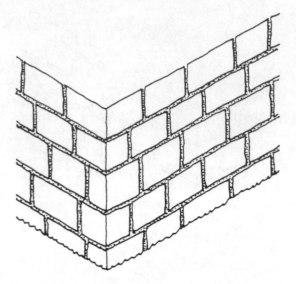

Squared rubble coursed

Fig. 113

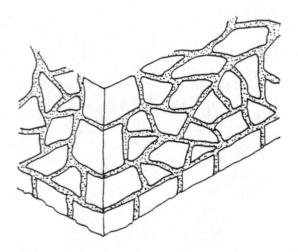

Polygonal walling

Fig. 114

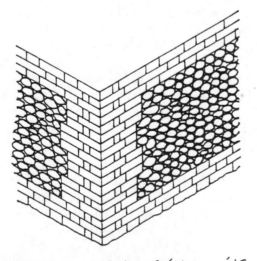

Uncoursed whole flints with brick dressing

Fig. 115

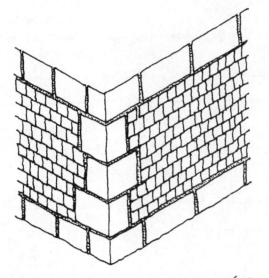

Squared knapped flints with
stone dressing

Fig. 116

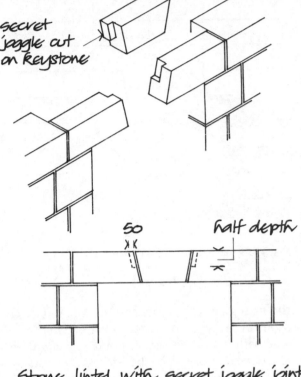

Stone lintel with secret joggle joints

Fig. 117

Openings in stone walls

Stone walls over door and window openings are supported by flat stone lintels or by segmental or semi-circular arches.

A stone lintel for small openings of up to about a metre wide can be formed of one whole stone with its ends built into jambs and its depth corresponding to one or more stone courses. The poor tensile strength of stone limits the span of single stone lintels unless they are to be disproportionately deep.

Over openings wider than about a metre it is usual to form lintels with three or five stones cut in the form of a flat arch. The stones are cut so that the joints between the ends of stones radiate from a common centre so that the centre or key stone, is wedge-shaped as illustrated in Fig. 117. The stones are cut so that the lower face of each stone occupies a third or a fifth of the width of the opening. To prevent the key stone sinking due to settlement and so breaking the line of the soffit, it is usual to cut half depth joggles in the ends of the key stone to fit to rebates cut in the other stones. The joggles and rebates may be cut the full thickness of each stone and show on the face of the lintel or more usually the joggles and rebates are cut on the inner half of the thickness of stones as secret joggles, which do not show on the face as

illustrated in Fig. 117. The depth of the lintel corresponds to a course height, with the ends of the lintel built in at jambs as end bearing. Stone lintels are used over both ashlar and rubble walling.

The use of lintels is limited to comparatively small openings due to the tendency of the stones to sink out of horizontal alignment. For wider openings some form of arch is used.

A stone arch consists of stones specially cut to a wedge shape so that the joints between stones radiate from a common centre, the soffit is arched and the stones bond in with the surrounding walling. The individual stones of the arch are termed 'voussoirs', the arched soffit the 'intrados' and the upper profile of the arch stones the 'extrados'.

Fig. 118 is an illustration of a stone arch whose soffit is a segment of a circle. The choice of the segment of a circle that is selected is to an extent a matter of taste, which is influenced by the appearance of strength. A shallow rise is often acceptable for small openings and a greater rise for larger, as the structural efficiency of the arch increases the more

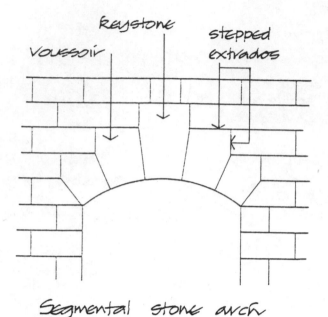

Segmental stone arch

Fig. 118

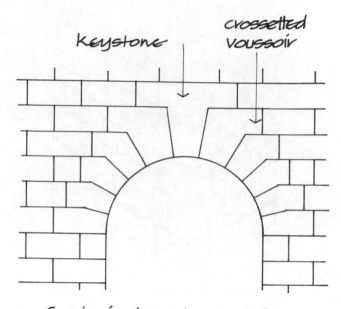

Semi-circular stone arch with crossetted voussoirs

Fig. 119

nearly the segment approaches a full half circle. The voussoirs of the segmental arch illustrated in Fig. 118 are cut with steps that correspond in height with stone courses, to which the stepped extrados is bonded.

The semi-circular stone arch illustrated in Fig. 119 has stones that are cut to form the arch and also to bond horizontally into the surrounding stone courses. The voussoirs are said to be crossetted. This extravagant cutting of stone is carried out purely for appearance sake. Most semi-circular arches have wedge-shaped voussoirs with stepped extrados similar to that of the segmental arch in Fig. 117.

Ashlar masonry joints: Ashlar stones may be finished with smooth faces and bedded with thin joints, or the stones may have their exposed edges cut to form a channelled or 'V' joint to emphasise the shape of each stone and give the wall a heavier, more permanent appearance. The ashlar stones of the lower floor of large buildings are often finished with channelled or V joints and the wall above with plain ashlar masonry to give the base of the wall an appearance of strength. Ashlar masonry finished with channelled or V joints is said to be rusticated.

Channelled joint (rebated joint) is formed by cutting a rebate on the top and one side edge of each stone, so that when the stones are laid, a channel rebate appears around each stone as illustrated in Fig. 120. The rebate is cut on the top edge of each

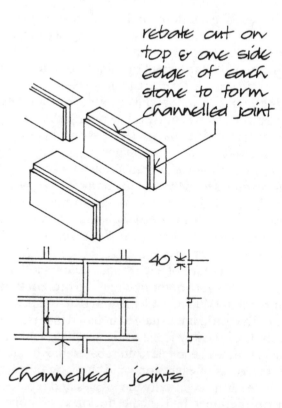

rebate cut on top & one side edge of each stone to form channelled joint

40

Channelled joints

Fig. 120

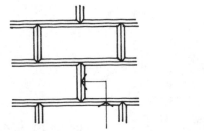

Vee-joints formed by cutting edges of stones at 45° degrees

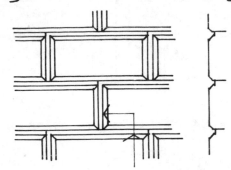

Vee and channelled joints formed by cutting rebate on edge of stones

Fig. 121

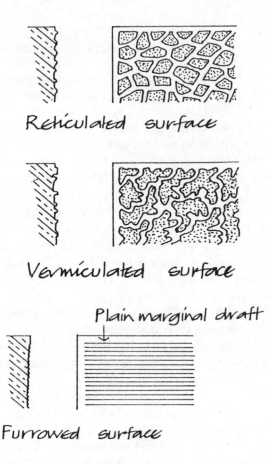

Reticulated surface

Vermiculated surface

Plain marginal draft

Furrowed surface

Chisel drafted margin

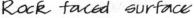

Rock faced surface

Plain marginal draft

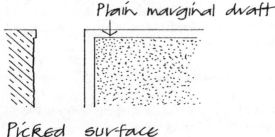

Picked surface

Fig. 122

stone so that when the stones are laid, rainwater which may run into the horizontal joint will not penetrate the mortar joint. V joint (chamfered joint) is formed by cutting the edges of stones so that when they are laid a V groove appears on face as illustrated in Fig. 121. Often the edges of stones are cut with both V and channelled joints as illustrated in Fig. 121.

Plain ashlar stones are usually finished with flat faces to form plain ashlar facing. The stones may also be finished with their exposed faces tooled to show the texture of the stone. Some of the tooled finishes used with masonry are illustrated in Fig. 122. It is the harder stones such as granite and hard sandstone that are more commonly finished with rock face, pitched face, reticulated or vermiculated faces. The softer, fine grained stones are usually finished as plain ashlar.

Resistance to weather and ground moisture

To prevent moisture rising from the ground through foundation walls it is necessary to form a continuous horizontal damp-proof course some 150 above ground level. One way of achieving this is to construct foundation walls of dense stone, such as granite, that does not readily absorb moisture. More usually one of the damp-proof materials described for use with brickwork is used. A sheet lead d.p.c. is commonly used as it is less likely to be squeezed out and forms a comparatively thin and therefore less unsightly joint than a bitumen felt d.p.c.

The resistance to the penetration of wind driven rain was not generally a consideration in the construction of solid masonry walls. The very considerable thickness of masonry walls of traditional large building was such that little, if any, rain penetrated to the inside face.

With the use of stone largely as a facing material for appearance sake, it is necessary to construct walls faced with stone as cavity walling with a brick or block inner leaf separated by a cavity from the stone faced outer leaf as illustrated in Fig. 108.

The outer leaf illustrated in Fig. 108 is built with natural stone blocks bonded to a brick backing, with full width stones in every other course and the stones finished on face as ashlar masonry. This is an expensive form of construction because of the considerable labour costs in preparing the ashlared stones. As alternatives the outer leaf of small buildings may be constructed with stone blocks by themselves for the full thickness of the outer leaf or with larger buildings the outer leaf may be constructed of brick to which a facing of stone slabs is fixed (see Volume 4).

The leaves of the cavity are tied with galvanised steel or stainless steel wall ties in the same way that brick and block walls are constructed and the cavity is continued around openings, or d.p.c.'s are formed to resist rain penetration at head, jambs and sills of openings.

Cornice and parapet walls

It is common practice to raise masonry walls above the level of the eaves of a roof, as a parapet. The purpose of the parapet is partly to obscure the roof and also to provide a depth of wall over the top of the upper windows for the sake of appearance in the proportion of the building as a whole. In order to

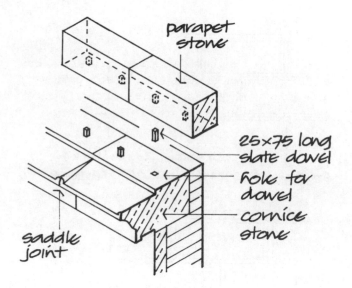

Dowel fixing for parapet stones

Fig. 123

provide a decorative termination to the wall a course of projecting moulded stones is formed. This projecting stone course is termed a cornice and it is generally formed some one or more courses of stone below the top of the parapet. Figure 123 is an illustration of a cornice and a parapet wall to an ashlar faced building. An advantage of the projecting cornice is that it affords some protection against rain to the wall below.

The parapet wall usually consists of two or three courses of stones capped with coping stones bedded on a d.p.c. of sheet metal. The parapet is usually at least $1B$ thick or of such thickness that its height above roof is limited by the requirements of The Building Regulations as described in Chapter 4 for parapet walls. The parapet may be built of solid stone or stones bonded to a brick backing.

The cornice is constructed of stones of about the same depth as the stones in the wall below, cut so that they project and are moulded for appearance sake. Because the stones project, their top surface is weathered (slopes out) to throw water off.

Saddle joint: The projecting, weathered top surface of coping stones is exposed and rain running off it will in time saturate the mortar in the vertical joints between the stones. To prevent rain soaking into these joints it is usual to cut the stones to form a saddle joint as illustrated in Fig. 123. The exposed top surface of the stones has to be cut to slope out (weathering) and when this cutting is executed a projecting quarter circle of stone is left on the ends of each stone. When the stones are laid, the projections on the ends of adjacent stones form a protruding semi-circular saddle joint which causes rain to run off away from the joints.

Weathering to cornices: Because cornices are exposed and liable to saturation by rain and possible damage by frost it is good practice to cover their exposed top surface with sheet metal. Obviously if the cornice is covered with sheet metal there is no point in having saddle joints formed. Sheet lead is usually preferred as a weathering because of its ductility and impermeability. Figure 124 is an illustration of lead weathering to a cornice.

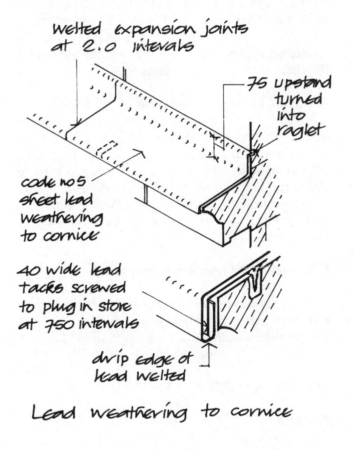

Lead weathering to cornice

Fig. 124

Cement joggles. Cornice stones project and one or more stones might in time settle slightly so that the decorative line of the mouldings cut on them would be broken and so ruin the appearance of the cornice. To prevent this possibility shallow V-shaped grooves

are cut in the ends of each stone so that when the stones are put together these matching V grooves form a square hole into which cement grout is run. When the cement hardens it forms a joggle which locks the stones in their correct position.

Dowels: To maintain stones in their correct position in a wall, slate dowels are used. The stones in a parapet are not kept in position by the weight of walling above and these stones are, therefore, usually fixed with slate dowels. These dowels consist of square pins of slate that are fitted to holes cut in adjacent stones as illustrated in Fig. 123.

Cramps: Coping stones are bedded on top of a parapet wall to prevent water soaking down into the wall below, but if the mortar in the joints between the coping stones cracks, rain will penetrate the cracks to the parapet below. If the parapet becomes saturated with rain it is possible that frost may damage it. So that rain cannot penetrate through cracks between coping stones and to keep the stones in correct alignment, it is usual practice to use cramps to strengthen the joints between the stones. These cramps consist either of a dovetailed piece of slate, Fig. 125, or a bronze cramp as illustrated in Fig. 126. As a barrier to the penetration of rainwater to the parapet it has become practice to bed the coping stones on a d.p.c. or to form a d.p.c. in the parapet wall at the level of the top of the upstand of a parapet gutter.

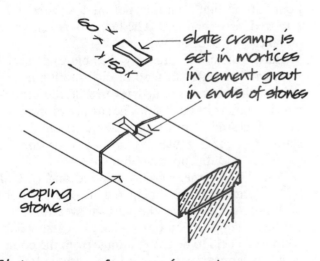

Slate cramp for coping stones

Fig. 125

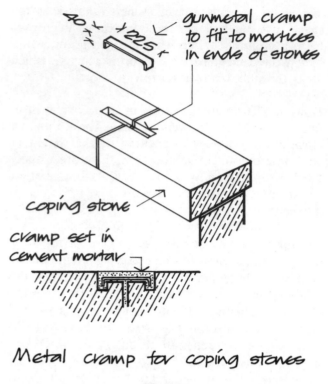

Metal cramp for coping stones

Fig. 126

Durability and freedom from maintenance

Sound natural stone is highly durable as a walling material and will have a useful life of very many years in buildings which are adequately maintained.

Granite is resistant to all usual weathering agents, including highly polluted atmospheres and will maintain a high natural polished surface for a hundred years or more. The lustrous polish will be enhanced by periodic washing.

Hard sandstones are very durable and inert to weathering agents but tend to dirt staining in time, due to the coarse grained texture of the material which retains dirt particles. The surface of sandstone may be cleaned from time to time to remove dirt stains by abrasive blasting with grit or chemical processes and thorough washing.

Sound limestone, sensibly selected and carefully laid is durable for the anticipated life of the majority of buildings. In time the surface weathers by a gradual change of colour over many years, which is commonly held to be an advantage from the point of view of appearance. Limestones are soluble in rainwater that contains carbon dioxide so that the surface of a limestone wall is to an extent self cleansing when freely washed by rain, while pro-

tected parts of the wall will collect and retain dirt. This effect gives the familiar black and white appearance of limestone masonry. The surface of limestone walls may be cleaned by washing with a water spray or by steam and brushing to remove dirt encrustations and the surface brought back to something near its original appearance.

In common with the other natural walling material, brick, a natural stone wall of sound stone sensibly laid will have a useful life of very many years and should require little maintenance other than occasional cleaning.

Fire resistance

Natural stone is incombustible and will not support or encourage the spread of flame. The requirements of Part B of Schedule 1 to The Building Regulations 1985 for structural stability and integrity and for concealed spaces apply to walls of stone as they do for walls of brick or block masonry.

Resistance to the passage of heat

The natural stones used for walling are poor insulators against the transfer of heat and will contribute little to thermal resistance in a wall. It is necessary to use some material with a low U value as cavity insulation in walls faced with stone in the same way that insulation is used in cavity walls of brick or blockwork.

Resistance to airborne and impact sound

Because natural building stone is dense it has good resistance to the transmission of airborne sound and will provide a ready path for impact sound.

TIMBER FRAMED WALLS

TIMBER

The word timber describes wood which has been cut for use in building. Timber has many advantages as a building material. It is a lightweight material that is easy to cut, shape and join by relatively cheap and simple hand or power operated tools in the production of either a single or a series of wall, floor and roof panels and frames, timber joist, stud, rafter and plate walls, floors and roofs and windows, doors and joinery generally. As a structural material it has favourable weight to cost, weight to strength and weight to modulus of elasticity ratios and coefficients of thermal expansion, K values, density and specific heat. With sensible selection, fabrication and fixing and adequate impregnation or protection it is a reasonably durable material in relation to the life of most buildings. Wood burns at temperatures of about 350°C and chars, the charred outer faces of the wood protecting the unburnt inner wood for periods adequate for escape during fires, in most buildings.

In this age of what the layman calls 'plastics' and 'synthetic' materials some people remark that timber is old-fashioned and suggest that more modern materials such as reinforced concrete should be used for floors and roofs. At present the cost of a timber upper floor for a house is about half that of a similar reinforced concrete floor and as the timber floor is quite adequate for its purpose it seems senseless to double the cost of the floor just to be what is called modern. Much of the timber used in buildings today is cut from the wood of what are called 'conifers'. A coniferous tree is one which has thin needle-like leaves that remain green all year round and whose fruit is carried in woody cones. Examples of this type of tree are fir and pine.

In North America and northern Europe are very extensive forests of coniferous trees, the wood from which can be economically cut and transported. The wood of coniferous trees is generally less hard than that of other sorts of trees, for example oak, and the wood from all conifers is classified as *soft wood*, whilst the wood from all other trees which have broad leaves is termed *hard wood*.

The cost of a typical soft wood used today is about half that of a typical hard wood.

Softwood is made up of many very thin long cells with their long axis along the length of the trunk or branch. These cells are known as tracheids and they give softwood its strength and texture, and serve to convey sap. The structure of hardwood is more complicated and in most hardwoods the bulk of the wood consists of long fibres. In addition to the fibres there are long cells, called vessels, which conduct water from the roots of the tree to its crown.

Trees whose wood is used for building are all *exogens*, meaning growing outwards, as each year most of these trees form new layers of wood under the bark. Beneath the protective bark around tree trunks and branches is a slimy light green skin. This is termed the cambium and it consists of a layer of wood cells which begin each spring to divide several times to form a layer of new wood cells. All these new wood cells are formed inside the cambium and the first wood cells formed each year are termed spring wood, and the cells formed later are termed summer wood. The wood cells which are formed first are thinner walled than those formed in the summer and in any cross section cut of wood there are distinct circular rings of light coloured spring wood then darker coloured and thinner rings of summer wood. As a new spring and summer ring is formed each year they are called annual rings, which are shown in Fig. 127.

In the diagram are shown a number of irregular radial lines marked medullary rays. These rays consist of specialised wood cells whose main purpose is to store food in readiness for it be conveyed to any part of the tree that may require it.

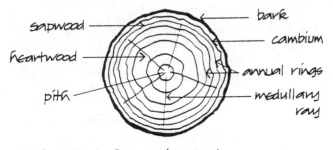

Section through growing wood

Fig. 127

Seasoning of timber

Up to two-thirds of the weight of growing wood is due to water in the cells of the wood. When the tree is felled and the wood is cut into timber this water begins to evaporate to the air around the timber, and the wood gradually shrinks as water is removed from the cell walls. As the shrinkage in timber is not uniform the timber may lose shape and it is said to warp. It is essential that before timber is used in buildings, either it should be stacked for a sufficient time in the open air for most of the water in it to dry out, or it should be artificially dried out. If wet timber is used in building it will dry out and shrink and cause cracking of plaster and twisting of doors and windows. The process of allowing, or causing, newly cut wood to dry out is called seasoning, and timber which is ready for use in building is said to have been properly seasoned.

Moisture content of timber: The amount of water in wood varies, and it is not sufficient to allow all timber to dry out for some specific length of time, as one piece of timber may be well seasoned and dried out, whilst another similar piece stacked for the same length of time may still be too wet to use immediately. It is necessary to specify that there shall be a certain amount of water, and no more or less, in timber suitable for building. It is said that timber shall have a certain moisture content, and the moisture content is stated as a percentage of the dry weight of the timber. The dry weight of any piece of timber is its weight after it is has been so dried that further drying causes it to lose no more weight. This dry weight is reasonably constant for a given cubic measure of each type of wood and is used as the constant against which the moisture content can be assessed. Table 14 sets out moisture contents for timber.

Table 14. Moisture Content of Timber for Categories of End Use

Position of timber in building	1 %	2 %
External uses fully exposed	18% or more	—
Covered and generally unheated	18%	24%
Covered and generally heated	16%	21%
Internal and continuously heated building	14%	19%

Column 1. Average moisture content attained in service conditions
Column 2. Moisture content which should not be exceeded at time of erection

Taken from BS5268:Part 2

The moisture content of timber should be such that the timber will not appreciably gain or lose moisture in the position in which it is fixed in a building.

Natural dry seasoning: When logs have been cut into timber it is stacked either in the open or in a rough open sided shed. The timbers are stacked with battens between them to allow air to circulate around them. The timbers are left stacked for a year or more, until most of the moisture in the wood has evaporated. Softwoods have to be stacked for a year or two before they are sufficiently dried out or seasoned,

and hardwoods for up to ten years. The least moisture content of timber that can be achieved by this method of seasoning is about 18%.

Artificial or kiln seasoning: Because of the great length of time required for natural dry seasoning and because sufficiently low moisture contents of wood cannot be achieved, artificial seasoning is largely used today. After the wood has been converted to timber it is stacked with battens between the timbers and they are then placed in an enclosed kiln. Air is blown through the kiln, the temperature and humidity of the air being regulated to effect seasoning more rapidly than with natural seasoning, but not so rapidly as to cause damage to the timber. If the timber is seasoned too quickly by this process it shrinks and is liable to crack and lose shape badly. To avoid this it is common practice to allow timber to season naturally for a time and then complete the process artificially as described.

Conversion of wood into timber

The method of cutting a log into timber will depend on the ultimate use of the timber.

Most large softwood logs are converted into timbers of different sizes so that there is the least wastage of wood. Smaller softwood logs are usually converted into a few long rectangular section timbers. Most hardwood today is converted into boards.

The method of converting wood to timber affects the timber in two ways: (a) by the change of shape of the timber during seasoning and (b) in the texture and differences in colour on the surface of the wood. Because the spring wood is less dense than the summer wood the shrinkage caused when wood is seasoned (dried) occurs mainly along the line of the annual rings. The circumferential shrinkage is greater than the radial shrinkage. Because of this the shrinkage of one piece of timber cut from a log may be quite different from that cut from another part of the log. This can be illustrated by showing what happens to the planks of a log converted by the 'through and through' cut method shown in Fig. 128. When the planks have been thoroughly seasoned their deformation due to shrinkage can be compared by putting them together in the order in which they were cut from the log as in Fig. 128. From this it will be seen that the plank which was cut with its long axis on the radius of the circle of the log, lost shape least

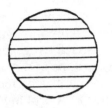 

Log through and through sawn   Deformation of planks due to shrinkage

Fig. 128

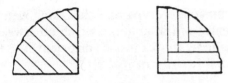

Log quartered & converted into boards

Fig. 129

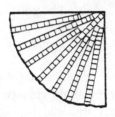

radial sawn boards

Fig. 130

noticeably and was the best timber after seasoning.

It is apparent that timber which is required to retain its shape during seasoning, such as good quality boarding, must somehow be cut as nearly as possible along the radius of the centre of the log. As it is not practicable to cut a log in the way we cut a slice out of an apple, logs which are to be cut along their radius are first cut into quarters. Each quarter of the log is then cut into boards or planks. This can be done in a variety of ways. Two of the most economical ways of doing this are shown here in Fig. 129. It will be seen from these diagrams that one or two boards or planks are cut very near a radius of the circle whilst the rest are cut somewhat off the radius and the former will lose shape least. In describing the structure and growth of a tree the medullary rays were described as being narrow radial lines of wood cells of different shape and structure than the main wood cell. If the face of a timber is cut on a radius of the circle of the log, the cells of the medullary rays may be exposed where the cut is made. With many woods this produces a very pleasing texture and colour on the surface of the wood and it is said that the 'figure' of the wood has been exposed. To expose the figure of wood by cutting along the medullary rays a quarter of a log has to be very wastefully cut as shown in Fig. 130. The radial cutting of boards as shown is very expensive and is employed only for high-class cabinet making and panelling timbers where the exposed figure of the wood will be used decoratively.

Decay in timber

Fungal decay: Any one of a number of wood-destroying fungi may attack timber that is persistently wet and has moisture content of over 20%.

Dry rot: This is the most serious form of fungal decay and is caused by *Seupula lacrymans* which can spread and cause extensive destruction of timber. The description dry rot derives from the fact that timber which has been attacked appears dry and powdery.

The airborne spores of this fungus settle on timber and if its moisture content is greater than 20% they germinate. (If the wood has less moisture content than 20% germination does not occur.) The spore forms long thread-like cells which pierce the wood cells and use the wood as a food. The thread-like cells multiply, spreading out long white thread-like arms called mycelium which feed on other wood cells. This fungus can spread many tens of feet from the point where the spore first began to thrive, and is capable of forming thick greyish strands which can find their way through lime mortar and the softer bricks.

Timber which is affected by this fungus turns dark brown and shrinks and dries into a cracked powdery dry mass which looks as though it has been charred by fire.

Prevention of dry rot: Never use unseasoned timber in buildings. Prevent seasoned timber becoming so wet that it can support the fungus by:

(a) building in a good horizontal damp-proof course;
(b) either ventilating the space below or around timber floors or by designing the building so that these spaces do not become damp;

(c) immediately repairing all leaking water, rain-water and drain pipes which otherwise might saturate timber to such an extent as to make it liable to dry rot.

Repair and renewal of timbers affected by dry rot: Every bit of timber which has been affected by the fungus or is in close proximity to it must be taken out or cut out of the building, and this timber should be burnt immediately. The purpose of burning the affected timber is to ensure that none of it is used in the repairs and to kill any spore that might cause further rot.

All walls on which, or against which, the fungus grew must be thoroughly cleaned and sterilised by application of a fungicidal solution. Any old lime plaster on which or through which the rot has spread should be hacked off and renewed in a cement plaster. New timber used to replace affected timber should be treated with a wood preservative before it is fixed or built in.

Wet rot is caused principally by *Coniophora puteana*, the cellar fungus, which occurs more frequently, but is less serious, than dry rot. Decay of timber due to wet rot is confined to timber that is in damp situations such as cellars, ground floors without damp-proof courses and roofs. The rot causes darkening and longitudinal cracking of timber and there is often little or no visible growth of fungus on the surface of timber.

Prevention of wet rot: Timber should not be built into or in contact with any part of the structure that is likely to remain damp. Damp-proof courses and damp-proof membranes above and at ground level and sensibly detailed flashings and gutters to roofs and chimneys will prevent the conditions suited to the growth of wet rot fungus.

Repair and renewal of timbers affected by wet rot: Affected timber must be cut out and replaced by sound new timber treated with a preservative. It is not necessary to sterilise brickwork around the area of affected timber.

Insect attack on wood

In this country the three sorts of insects which most commonly cause damage to timber are the furniture beetle, the death-watch beetle and the house long-horn beetle.

The common furniture beetle *(Anobium punctatum)* is the beetle whose larvae most commonly attack timber and furniture and its attack is generally known as 'wood worm'. Between June and August the beetles fly around and after mating the female lays small eggs in the cracks and crevices of any dry softwood or hardwood. The eggs hatch and a small white grub emerges. These grubs (larvae) have powerful biting jaws with which they tunnel into the wood, using the wood as food. The grubs tunnel along the grain of the wood for a year or two until they are fully grown, then they bore to just below the surface of the wood where they pupate or, change into beetles. These emerge in June to August and the life cycle of the beetle is repeated. The hole which each grub makes is very small and timber is structu-rally weakened only if a great number of these holes tunnelled by a number of grubs are formed over many years. The holes are very unsightly, particu-larly if the beetle infests furniture or panelling.

The death-watch beetle *(Xestobium rufovillo-sum):* The grubs of this beetle thrive particularly well on very old timbers that have suffered decay and rarely attack softwoods. The beetle lays eggs which hatch into grubs (larvae) which, in turn, do as the larvae of the furniture beetle, tunnel along the wood for from one to several years. The name death-watch beetle derives from the ticking sound made by the beetles as they tap their heads on timber during the mating season in May and June. Again only after years of infestation by this beetle is the strength of timber seriously affected.

House longhorn beetle *(Hylotrupes bajulus):* A large insect that has in recent years been active in London and the home counties, that attacks mainly soft-wood. It has a life cycle of up to 11 years and can cause such extensive damage that only a thin shell of sound timber is left. There is little external evidence of the infestation except some unevenness of wood surfaces due to borings just below the surface. The Building Regulations 1976 specified areas of the south of England in which the softwood timbers of roofs are required to be treated with a suitable preservative against attack by the longhorn beetle.

The powder-post beetle: The larvae of this beetle only attack hardwoods stored in timber yards. The name derives from the powder found at the foot of hardwood posts stored in timber yards.

Control of attacks by beetles: Timber which has been affected by the larvae of these beetles should be sprayed or painted with a proprietary preservative during early summer and autumn. These preservatives prevent the larvae changing to beetles at the surface of the wood and so arrest further infestation.

Wood preservation

Timber is treated with preservatives to protect it against damage by fungi and insects. Wood preservatives are divided into three main groups:

(a) tar oils
(b) water-borne
(c) organic solvent.

Tar oil preservatives are blends of distillate oils of coal tar, commonly known as creosote. Bethell's Patent for coal tar creosote (taken out in 1838) is still widely used today as a preservative for railway sleepers, posts, farm buildings and fences. Creosote reduces moisture absorption and is a very effective protection against fungal and insect attack. It is a brown viscous liquid which has a strong odour and cannot be painted over. It is extensively used to preserve timbers used externally, where appearance is not a consideration.

Timber should be adequately seasoned before treatment with a preservative. Creosote may be painted and sprayed on surfaces to give protection to the depth of the penetration of the liquid below surface. Timbers steeped or immersed in creosote are protected to the extent of the penetration of the creosote which is a product of the nature of the wood and the length of time of immersion. Pressure impregnation is the most effective method of preservation. There are two methods of impregnation, the full cell and the empty cell process. In the full cell process timbers are sealed in a vacuum cylinder in which air is removed from the wood cells which are then filled with creosote under pressure to give full protection of all the cells of the timber. In the empty cell process creosote is introduced under pressure after the timbers have been subjected to pressure and then surplus creosote is withdrawn by vacuum so that all the wood cells are given adequate protection. This latter process, which uses less creosote, is less likely to cause bleeding of creosote from the surface than the full cell process.

Water-borne preservatives: The two types of preservative used, boron compounds for green (not seasoned) timber and combinations of copper, chrome and arsenic (C.C.A.) for seasoned timber are both water soluble. The C.C.A. preservatives are usually used to pressure impregnate timbers which after treatment have high resistance to attack by fungi and insects. The C.C.A. water-borne preservative is commonly used for structural timbers liable to attack. The preservative gives light coloured wood a light grey/green colour and is liable to cause some slight swelling of timber.

Organic solvent type preservatives: These preservatives contain fungicidal and/or insecticidal ingredients of low water solubility that are dissolved in volatile solvents such as white spirit that evaporate, leaving the fungicide and insecticide in the wood. These preservatives do not cause any change in moisture content or dimension in wood and are particularly suitable for joinery timbers. They are non-corrosive to metals and can be satisfactorily over-painted. Pressure impregnation is the most effective method of using these preservatives.

Finishes for timber

There are three types of finish for wood, paint, varnish and stains. The traditional finishes paint and varnish are protective and decorative finishes which afford some protection against water externally and provide a decorative finish which can easily be cleaned internally. Paints are opaque and hide the surfaces of the wood whereas varnishes are sufficiently transparent for the texture and grain of the wood to show. Of recent years stains have been much used on timber externally. There is a wide range of stains available, from those that leave a definite film on the surface to those that penetrate the surface and range from gloss through semi-gloss to matt finish. The purpose of this finish is to give a selected uniform colour to wood without masking the grain and texture of the wood. Most stains contain a preservative to inhibit fungal surface growth. These stains are most effective on rough sawn timbers.

TIMBERS WALLS

The construction of a timber framed wall is a rapid, clean, dry operation. The timbers can be cut and assembled with simple hand or power operated tools and once the wall is raised into position and fixed it is ready to receive wall finishes. A timber framed wall has adequate stability and strength to support the floors and roof of small buildings, such as houses. Covered with wall finishes it has sufficient resistance to damage by fire, good thermal insulating properties and reasonable durability providing it is sensibly constructed and protected from decay. In North America timber is as commonly used for walls as brick is in the United Kingdom. A timber framed house can be constructed on site by two men in a matter of a few days.

There has been a prejudice against timber buildings in this country for many years. For some years after the end of the Second World War (1945) timber framed houses were built as a means of satisfying the need for new housing, by the rapid building possible with this form of construction. Comparatively few such houses were built partly because of an inherent feeling that timber was not as solid or durable a material as brick and through the unwillingness of building societies to lend money for the construction or purchase of these houses.

More recently timber framed houses, generally with brick or block outer leaves, have been constructed and once again the timber wall frame has been given a bad name because of problems of condensation in the fabric of walls due to ill-considered design.

It is irrational to construct a timber frame that has by itself adequate strength to support the floors and roof of a house and then enclose it with a brick or block wall purely for the sake of appearance. The combination of these two systems of construction impedes the rapid construction possible with timber and confuses the 'wet trades' system of the construction of solid walling with that of the 'dry system' of construction of timber framed construction.

Strength and stability

Strength: The strength of timber varies with species and is generally greater with dense hardwoods than less dense softwoods. Strength is also affected by defects in timber such as knots, shakes, wane and slope of the grain of the wood.

Stress grading of timber

There is an appreciable variation in the actual strength of similar pieces of timber which had led in the past to very conservative design in the use of timber as a structural material. Uncertain of the strength of individual timbers, it was practice to overdesign, that is, make allowance for the possibility of weakness in a timber and so select timbers larger than necessary.

Of recent years, systems of stress grading have been adopted with the result that a more certain design approach is possible with a reduction of up to 25% in the section of structural timbers. Stress grading of structural timbers, which was first adopted in The Building Regulations 1972, is now generally accepted in selecting building timber.

There are two methods of stress grading: visual grading and machine grading.

Visual grading: Trained graders determine the grade of a timber by a visual examination from which they assess the effect on strength of observed defects such as knots, shakes, wane and slope of grain. There are two visual grades, General Structural (GS) and Special Structural (SS), the allowable stress in SS being higher than in GS.

Machine grading: Timbers are subjected to a test for stiffness by measuring deflection under load in a machine which applies a specified load across overlapping metre lengths to determine the stress grade. This mechanical test, which is based on the fact that strength is proportional to stiffness, is a more certain assessment of the true strength of a timber than a visual test. The machine grades, which are comparable to the visual grades, are Machine General Structural (MGS) and Machine Special Structural (MSS). There are in addition two further machine grades, M50 and M75.

Stress graded timbers are marked GS and SS at least once within the length of each piece for visually graded timber together with a mark to indicate the grader or company. Machine graded timber is likewise marked MGS, M50, MSS and M75 together with the BS kitemark and the number of the British Standard, 4978.

Approved Document A, which gives practical guidance to meeting the requirements of Schedule 1 to The Building Regulations 1985 for small buildings, includes tables of the sizes of timber required for floors and roofs, related to load and span.

Stability: The stability of a timber framed wall depends on a reasonably firm, stable foundation on which a stable structure is to be constructed. In common with other systems of walling the foundation to a timber framed wall should serve as a firm, level base that will transmit the loads of the building to the ground without undue settlement or movement. The strip or raft foundations described for brick and block walls will serve equally for the timber framed walls of small buildings, depending on ground conditions. Because timber framing is a lightweight form of construction it will depend less on support from the foundation than will a similar brick construction and more on being firmly anchored to the foundation against uplift due to wind forces. It will be seen from Fig. 131 that the base of the timber framed wall is bolted to the supporting brick foundation wall for this reason.

A brick foundation on strip foundation or a concrete curb upstand on a raft or slab foundation is usual with the soleplate bedded some 150 above ground level.

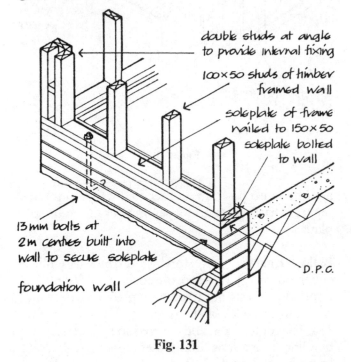

double studs at angle to provide internal fixing

100 × 50 studs of timber framed wall

soleplate of frame nailed to 150 × 50 soleplate bolted to wall

13 mm bolts at 2 m centres built into wall to secure soleplate

foundation wall

D.P.C.

Fig. 131

The stability of a timber framed wall depends on the size of timbers used and the method of framing timber sections as wall and roof elements of structure to provide adequate strength and rigidity.

Timber framed walls consist of small section timbers fixed vertically at centres to suit the loads to be supported and materials to be used as weathering and facing, fixed to a bottom and a top member to form the traditional timber stud frame illustrated in Fig. 132, which derives its name from the vertical members called studs. The bottom and top members, termed sole (or cill) and head plate respectively, are of the same section of timber as the studs. Figure 132 is an illustration of a typical timber stud frame suitable for both external and internal walls and load bearing and non-load bearing walls. Fixed between load bearing cross walls as the front or back wall of a terrace house or fixed internally between brick or block walls, this simple stud frame has adequate stability and needs no additional bracing.

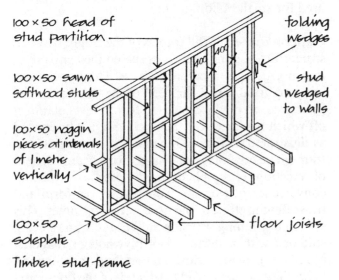

100 × 50 head of stud partition

folding wedges

100 × 50 sawn softwood studs

stud wedged to walls

100 × 50 noggin pieces at intervals of 1 metre vertically

100 × 50 soleplate

floor joists

Timber stud frame

Fig. 132

The simple timber stud wall by itself, with the vertical studs nailed to the sole and head plates, does not have sufficient stability or rigidity to be used for the walls of buildings because the non-rigid connection of the members of the frame will not strongly resist racking along the length of the frame due to lateral forces such as wind and eccentric loading. The timber stud wall has, therefore, to be braced against racking if it is to be used as a freestanding timber frame. Longitudinal bracing may be provided by diagonal timber braces built into the frame or a wall covering of diagonal boarding or plywood can be used to provide sufficient rigidity. Practice today is to use the sheathing of boards or plywood as bracing.

Because of its small mass, a timber frame wall has poor lateral stability against lateral forces, such as wind, that tend to overturn the wall. For lateral stability a timber frame wall is buttressed by the connection of external and internal walls or partitions that are connected at corners and internal junctions in a building. Upper floors and roofs that bear on the timber frame wall also provide some lateral stability up the height of walls.

The small lightweight sections of timber used lend themselves to a system of prefabrication either on site or off. Practice is to assemble stud frames on a floor or bench and then to manhandle them into position on site. The simplest way of assembling or prefabricating timber stud walls is to construct wall or storey height frames that can be raised into position without heavy lifting gear.

The three systems of timber wall framing that are used for small buildings are:

Platform frame: Wall frames of the height between structural floor and ceilings are assembled and raised into position on a prepared base. The upper floor joists are fixed across the head of the frames and the floor is covered with boards to form a level platform off which the frames of the next floor can be erected as illustrated in Fig. 133, hence the name platform frame. The advantage of this platform frame system of erection is simplicity and speed of erection, convenience in working from a level platform and the stiffening effect of the floor between frames. This system of framing depends on the stud frames being stiffened with a sheathing of plywood or diagonal boards to provide rigidity in handling, erection and in use. The header nailed to the ends of the floor joists serves to seal the cavities between joists for insulation and as a fixing for external finishes.

Storey frames: Wall frames of storey height, that is the height between floors, are assembled, raised and fixed to a prepared base. Upper floor joists are fixed to a plate fixed to the studs and the next wall frame is raised and fixed directly on top of the lower frame as illustrated in Fig. 133. This system of stud framing and assembly is particularly suited to frames that are assembled with the external weathering of boards in place so that either an overlap or a weathering strip will complete the joint between frames.

The advantages of the platform and storey frames are flexibility in positioning openings between one storey and another, the comparatively short lengths of timber minimise warping due to drying shrinkage and these lightweight panels can be manhandled. A timber framed wall may be assembled in two or more panels for ease of handling and erection. The panels

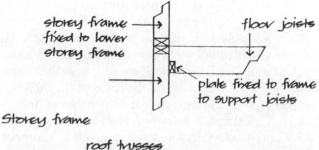

Storey frame

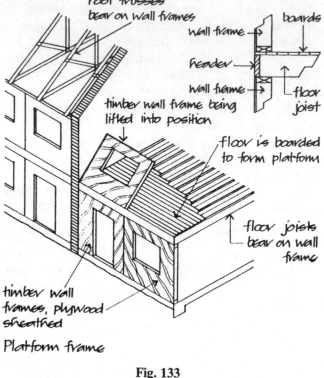

Platform frame

Fig. 133

are lifted into position and secured to a continuous separate head plate and a separate continuous sole plate.

Balloon frame is a two storey stud frame assembled either on or off site and raised into position. The upper floor joists are fixed on a plate or ledger and nailed to the sides of the studs as illustrated in Fig. 134. The roof is fixed and covered so that subsequent building operations are under cover. A balloon frame panel is particularly suited for use as a side wall of a two storey timber framed building with the end walls of platform or storey frames to support the upper floor and roof. The disadvantage of a balloon frame is its size for transport and handling, the long lengths of stud which may twist due to drying shrinkage and the limitation on choice of position of openings due to the continuous studs.

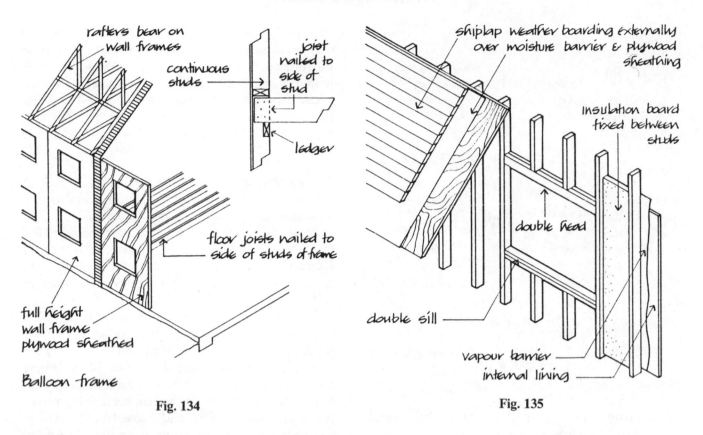

Fig. 134

Fig. 135

Foundation: Timber wall frames are erected on a foundation of conventional walls or a concrete base or raft to which a sole plate is bolted as illustrated in Fig. 131. The sole plate is bedded on a d.p.c. Where a wall is assembled as one frame, the sole of the frame may be bolted direct to the foundation. Where the wall is made up of two or more frames then a separate sole plate is first bolted to the foundation and the separate frames are nailed to it.

Angles and intersections: To provide a secure fixing for internal and external finishes and as a means of fixing frames at angles and intersections for stability, it is usual to use an additional stud as illustrated in Fig. 131.

Openings for windows and doors are framed with either single or double head and cill timbers depending on the width of the opening, single members for narrow and double for wide as illustrated in Fig. 135.

Resistance to weather and ground moisture

To prevent moisture rising from the ground through the foundation to a timber framed wall the sole plate of walls is fixed some 150 above ground and formed

on a continuous horizontal d.p.c. on one of the bitumen felt materials described for brick and block walls.

By itself a timber wall frame has little resistance to rain penetration. Where the frame is constructed without an external leaf of brick or block it is covered with a sheathing of timber boards or plywood that serves as a bracing along the length of the frame and also as a first sheath against wind and rain and as a background for breather paper and the weather-board or tile or slate or other cladding that serves to resist the penetration of rain. The traditional weather envelope for timber framed walls is timber weather-boarding nailed horizontally across the stud frame as illustrated in Fig. 136. The boards are fixed to overlap down the wall to shed water. In exposed situations a cladding of tile or slate nailed to battens would be used. More recently profiled sheets of metal or plastic have been used as a weathering with none too happy results as regards appearance as the large profile of these sheets is out of scale to small buildings and because of the difficulties of trimming the sheets around openings.

An external leaf of brick or block will at once provide the sense of permanence for which it is used and at the same time act as a weather shield against

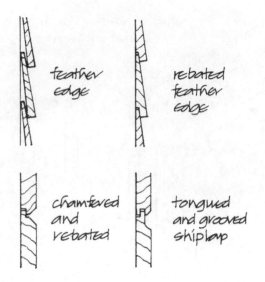

timber weatherboarding

Fig. 136

wind driven rain by virtue of the cavity formed between the two leaves.

Durability and freedom from maintenance

Sound, well seasoned timber protected from insect and fungal attack and rainwater and moisture vapour will have a useful life of very many years and should require no maintenance. Softwood weatherboarding and windows and doors will need periodic maintenance and renewal of protective coatings.

Resistance to fire

The surface of timber that is exposed to fire ignites and burns at comparatively low temperatures and chars to form a surface of charcoal which insulates the wood below and delays ignition as the temperature rises. In fires, the exposed surface of large timbers will ignite and char and the good thermal insulation of wood will prevent a significant rise in the temperature of the timber not exposed to fire and so delay ignition and burning of timber for some time. As there is little loss of strength with rise in temperature, large timbers may maintain their structural stability for some time.

Because timber ignites at comparatively low temperatures there is significant spread of flame over the surface of the material. There are flame retardant materials that will reduce surface spread of flame. These flame retardant materials are applied either by pressure impregnation of water soluble salts or resins or as surface coatings such as intumescent phosphate resin.

Concealed spaces: A requirement from Part B of Schedule 1 to The Building Regulations 1985 is that concealed spaces (cavities) shall be sealed and subdivided where necessary to inhibit unseen spread of fire and smoke. The practical guidance in Approved Document B requires barriers to the edges of all cavities and around all openings in the cavities in walls.

The cavities in a timber framed wall that should be sealed are the cavity between brick outer and timber framed inner leaves of a cavity wall and the space between a timber wall and external cladding. Where the cavity in a timber framed wall meets the cavity in floors and roofs a continuous seal must be formed to prevent the spread of fire and smoke from one cavity to another. It is also necessary to form seals around all door and window openings in these walls.

The materials most used as cavity barriers or fire stops, in timber framed walls are mineral wool at least 50 thick and timber at least 38 thick.

The two forms of mineral wool cavity barrier that are used are wire mesh reinforced mineral wool blanket and mineral wool sleeved cavity barrier formed as a polythene sleeve around mineral wool. The mineral wool blanket, reinforced with wire mesh, is for use principally as a cavity barrier at the horizontal and vertical junctions of cavities in timber walls, floors and roofs and the junction of timber framed external walls with internal walls of timber. The wire mesh is stapled to the timber frame and the blanket is doubled and pressed into the cavity. The polythene sleeved mineral wool barrier is designed for use in the cavity of external walls with brick outer leaves and timber inner leaves. The polythene sleeve acts as a barrier to moisture that might otherwise bridge across the cavity through the mineral wool. The polythene sleeve has two flanges that served as fixing by stapling to the timber frame around openings and at horizontal and vertical junctions of cavities in walls with floors and roof. The sleeved barriers should be pressed firmly into place to fill the cavity. Timber headers and battens at least 38 thick that completely and continuously fill the cavity may also be used as cavity barriers.

Resistance to the passage of heat

Timber is a comparatively good insulator having a U value of 0.13 for softwood and 0.15 for hardwood. The sections of a timber frame do not by themselves afford sufficient insulation to meet the requirements of the Building Regulations 1985 and a layer of some insulating material has to be incorporated in the construction.

The layer of insulation is fixed either between the vertical studs of the frame or on the outside face of the framing over or under the plywood or board sheathing or it may be fixed both between the studs and on the outside face.

The disadvantage of fixing the insulation between the studs is that there will be a deal of wasteful cutting of insulation boards to fit them between studs and to the extent that the U value of the timber stud is less than that of the insulation material, there will be a degree of thermal bridge across the studs.

The advantage of fixing the insulation across the outer face of the timber frame is simplicity in fixing and the least amount of wasteful cutting and that the void space between the studs will augment insulation and provide space in which to conceal service pipes and cables. Where the timber wall serves as an inner leaf to a cavity with an outer leaf of brick or block, it is usual to fix the insulation on the outside face of the timber frame over the plywood or board sheathing as illustrated in Fig. 137.

Insulation between studs in addition to insulation on the outside face of a timber frame is used to augment the outer insulation and also as a form of acoustic insulation against airborne sound where the comparatively lightweight timber wall is the only barrier to sound. The insulation between studs is usually one of the mineral fibre boards.

Vapour check: The high level of insulation required for walls may well encourage moisture vapour held by warm inside air, to find its way due to moisture vapour pressure, into a timber framed wall and condense to water on the cold side of the insulation. The condensation moisture may then damage the timber frame. As a barrier to warm moist air there should be some form of vapour check fixed on the warm side of the insulation. Closed cell insulating materials such as extruded polystyrene, in the form

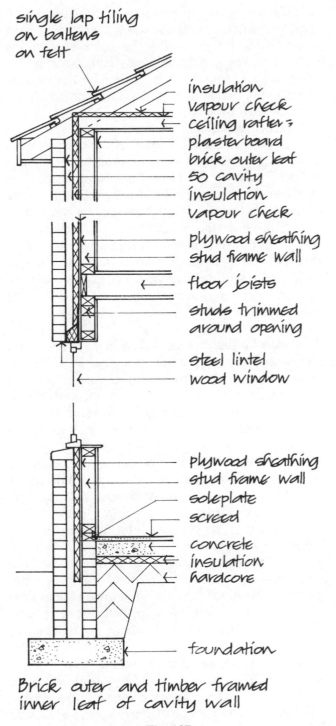

single lap tiling on battens on felt

insulation
vapour check
ceiling rafters
plasterboard
brick outer leaf
50 cavity
insulation
vapour check

plywood sheathing
stud frame wall

floor joists

studs trimmed around opening

steel lintel
wood window

plywood sheathing
stud frame wall
soleplate
screed
concrete
insulation
hardcore

foundation

Brick outer and timber framed inner leaf of cavity wall

Fig. 137

of rigid boards, are impermeable to moisture vapour and will by themselves act as a vapour check. The boards should either be closely butted together or supplied with rebated or tongued and grooved edges so that they fit tightly and serve as an efficient vapour check. Where insulation materials that are pervious to moisture vapour, such as mineral fibre boards are used for insulation a vapour check of polythene sheet must be fixed right across the warm side of the insulation. The polythene sheet should be lapped at joints and continued up to unite with any vapour check in the roof and should, as far as practical, not be punctured by service pipes.

Electrical cables that are run through the members of a timber wall and the insulation between the studs, may overheat due to the surrounding insulation with a risk of short circuit or fire. To prevent overheating of cables run through insulation, the cables should be derated by a factor of 0.75 by using larger cables than specified, which will generate less heat. So that cables are not run through insulation it is wise to fix the inside dry lining to timber frames that are filled with insulation, on timber battens nailed across the frame so that there is a void space in which cables can be safely run.

Insulation for timber walls: The layer of insulation to a timber wall by itself or to a cavity wall with brick or block and timber framed leaves is fixed between the studs or on to the outside face of the timber wall.

The inorganic materials glass fibre and rockwool are most used for insulation between studs as there is no advantage in using the more expensive organic materials, as the thickness of insulation required is not usually greater than the width of the studs. Either rolls of loosely felted fibres or compressed semi-rigid batts or slabs of glass fibre or rockwool are used. The material in the form of rolls is hung between the studs where it is suspended by top fixing and a loose friction fit between studs, which generally maintains the insulating material in position for the comparatively small floor heights of domestic buildings. The friction fit of semi-rigid slabs or batts between studs, is generally sufficient to maintain them, close butted, in position.

For insulating lining to the outside face of studs, either inside a cavity or on the outside face of a timber wall behind cladding, one of the organic insulants such as XPS or PIR provide the advantage of least thickness of insulating material for given resistance to the transfer of heat. The more expensive organic insulants, in the form of boards, are fixed across the face of studs for ease of fixing and to save wasteful cutting.

A vapour check should be fixed on to or next to the warm inside face of insulants against penetration of moisture vapour. Organic insulants, such as XPS,

which are substantially impervious to moisture vapour can serve as a vapour check particularly when rebated edge boards are used and the boards are close butted together.

Table 15 lists some of the insulants suited for use in timber framed walls.

Resistance to airborne and impact sound

The small mass of a timber framed wall affords little resistance to airborne sound and does not readily conduct impact sound. The insulation necessary for the conservation of heat will give some reduction in airborne sound and the use of a brick or block outer leaf will appreciably reduce the intrusion of airborne sound.

Table 15. Insulating Materials

Timber framed wall (insulation between studs)	Thickness	U valve W/m²K
Glass fibre rolls	50, 80, 90, 100	0.04
semi-rigid batts	80, 90, 100, 120, 140, 160	0.04
Rockwool rolls	60, 80, 90, 100, 150	0.037
semi-rigid slabs	60, 80, 90, 100	0.037
Timber framed wall (insulation fixed to face of studs)		
XPS boards T & G on long edges	25, 50	0.025
PIR boards with heavy duty aluminium facings	20, 25, 30, 35, 50	0.02

XPS extruded polystyrene
PIR rigid polyisocyanurate

CHAPTER THREE

FLOORS

FUNCTIONAL REQUIREMENTS

The functional requirements of a floor are:

Strength and stability
Resistance to weather and ground moisture
Durability and freedom from maintenance
Fire resistance
Resistance to the passage of heat
Resistance to airborne and impact sound

Strength

The strength of a floor depends on the characteristics of the materials used for the structure of the floor, such as timber, steel or concrete. The floor structure must be strong enough to safely support the dead load of the floor and its finishes, fixtures, partitions and services and the anticipated imposed loads. Dead loads are calculated from the unit weight of the materials set out in BS 648 and imposed loads from BS 6399: Part 1. From these calculations the required strength of the floor is determined. Where imposed loads are small, as in single family domestic buildings of not more than 3 storeys, the timber floor construction may be determined from the span of the floor and the imposed loads set out in the tables for joists in Approved Document A, as practical guidance to meeting the requirements of Schedule 1 to The Building Regulations 1985. The two tables for floor joists, for two timber strength classes, relate dead load and spacing of joists and size of joist to spans.

Stability

A floor is designed and constructed to serve as a horizontal surface to support people and their furniture, equipment or machinery. The floor should have adequate stiffness to remain reasonably stable and horizontal under the dead load of the floor structure and such partitions and other fixtures it supports and the anticipated static and live loads it is designed to support. The floor structure should also support and accommodate either in its depth, or below or above, electrical, water, heating and ventilating services without affecting its stability. For stability there should be adequate support for the floor structure and the floor should have adequate stiffness against gross deflection under load.

Solid ground and basement floors are usually built off the ground from which they derive support. The stability of such floors depends, therefore, on the characteristics of the hardcore and soil under them. For small domestic loads the site concrete, without reinforcement, provides adequate stability. For heavier loads, such as heavy equipment or machinery, a reinforced concrete slab is generally necessary with, in addition, a separate pile foundation under heavy machinery.

On shrinkable clay soils it may be necessary to use a suspended reinforced concrete slab against differential expansion or contraction of the soil, especially where there are deep rooted trees near the building.

Upper or suspended floors are supported by walls or beams and should have adequate stiffness to minimise deflection under load. Under load a floor will deflect and bend and this deflection or bending should be limited to avoid cracking of rigid finishes such as plasterboard and to avoid the sense of apprehension in those below the floor that they might suffer, if the deflection or bending were obvious. A deflection of about 1/300 of the span is generally accepted as a maximum in the design of floors.

Resistance to weather and ground moisture

The ground floor of a building, especially a heated building, will tend to encourage moisture from the ground below to rise and make the floor damp and feel cold and uncomfortable. This in turn may require additional heating to provide reasonable conditions of comfort. An appreciable transfer of moisture from the ground to the floor may promote conditions favourable to wood rot and so cause damage to timber ground floors and finishes.

Obviously the degree of penetration of moisture from the ground to a floor will depend on the nature of the subsoil, the water table and whether the site is level or sloping. On a gravel or coarse grained sand base, where the water table throughout the year is well below the surface, there will be little penetration whereas on a clay base, with the water table close to the surface, there will be appreciable penetration of moisture from the ground to floors. In the former instance a concrete slab alone may be a sufficient barrier and in the latter a waterproof membrane on, in or under the concrete slab will be necessary to prevent moisture rising to the surface of the floor. The requirements from Part C of Schedule 1 to The Building Regulations 1985 for the resistance of the passage of moisture to the inside of buildings are described in Chapter 1.

Durability and freedom from maintenance

Ground floors on a solid base protected against rising moisture from the ground, and suspended upper floors solidly supported and adequately constructed and protected inside a sound envelope of walls and roof, should be durable for the expected life of the building and require little maintenance or repair.

Fire resistance

Suspended upper floors should be so constructed to provide resistance to fire for a period adequate for the escape of the occupants from the building. The notional periods of resistance to fire, from $\frac{1}{2}$ to 4 hours, depending on the size and use of the building, are set out in the Building Regulations. In general a timber floor provides a lesser period of resistance to fire than a reinforced concrete floor. In consequence timber floors will provide adequate resistance to fire in small domestic buildings, and concrete floors the longer periods of resistance to fire required in large buildings, where required periods of resistance are higher.

Resistance to the passage of heat

A floor should provide resistance to transfer of heat where there is normally a significant air temperature difference on the opposite sides of the floor, as, for example, where a floor is exposed to outside air. Where an open car port is formed under a building and the floor over the port is exposed to outside air, the floor over should be insulated and have a U value the same as an exposed wall.

Obviously a ground floor should be constructed to minimise transfer of heat from the building to the ground or the ground to the building. Both hardcore and a damp-proof membrane on, under or sandwiched in the oversite concrete will assist in preventing the floor being damp and feeling cold and so reduce heating required for comfort and reduce transfer of heat. The use of insulation under solid ground floors is described in Chapter 1.

Where under floor heating is used it is essential to introduce a layer of insulation below and around the edges of the floor slab to reduce transfer of heat to the ground.

Resistance to airborne and impact sound

Upper floors that separate dwellings, or separate noisy from quiet activities, should act as a barrier to the transmission of airborne sound and reduce impact sound. The comparatively low mass of a timber floor will transmit airborne sound more readily than a high mass concrete floor, so that floors between dwellings, for example, are generally constructed of concrete. The resistance to sound transmission of a timber floor can be improved by filling the spaces between the timber joists with either lightweight insulating material or a dense material. The additional cost of such filling to a new floor for the comparatively small reduction in sound transmission may not be worthwhile where, for a modest increase in cost, a concrete floor will be more effective. Where existing buildings, with timber floors, are to be converted into flats the only reasonable way of improving sound insulation between floors is the use of filling between joists and some form of floating floor.

The reduction of impact sound is best effected by a floor covering such as carpet or a resilient layer under the floor surface, that deadens the sound of footsteps on either a timber or a concrete floor.

The hard surfaces of the floor and ceiling of both timber and concrete floors will not appreciably absorb airborne sounds which will be reflected and may build up to an uncomfortable level. The sound absorption of a floor can be improved by carpet or felt, and the ceiling by the use of one of the absorbent 'acoustic' tile or panel finishes.

CONCRETE GROUND FLOORS

Ground supported slab

The majority of ground floors are constructed as ground supported in-situ cast concrete slabs on a hardcore bed with a damp-proof membrane and insulation as described in Chapter 1.

Strength and stability

Suspended concrete slabs: Where the ground under a floor has poor or uncertain bearing capacity, or is liable to volume change due to seasonal loss or gain of moisture and a ground supported slab might sink or crack due to settlement, it is wise to form the ground floor as a suspended reinforced slab supported by external and internal load bearing walls, independent of the ground.

Suspended concrete slabs are constructed with one of the pre-cast reinforced concrete plank, slab or beam and block floor systems described later for upper floors, because there is no ready means of constructing centering on which to cast an in-situ concrete floor. The one way spanning, pre-cast concrete floor bears on internal and external load bearing foundation walls with end bearing of at least 90 and is built into the walls. The depth of the plank, slab or beams depends on the loads to be carried and the span between supporting walls.

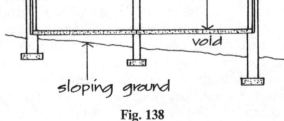

Fig. 138

Resistance to ground moisture

Damp-proof membrane: Where the suspended ground floor slab is formed above the reduced ground level inside the building, with an air space of at least 75 below the underside of the slab and the air space is ventilated to outside air, then it is not necessary to use a d.p.m. The purpose of ventilating the space below a suspended floor is to prevent the build up of stagnant moist air in the space, which would otherwise tend to make the slab damp. When the ground inside a building with an air space below a suspended slab, is below the lowest level of the surrounding ground or due to the slope of the land is liable to collect water that is not drained, then a d.p.m. should be formed as illustrated in Fig. 138.

Resistance to the passage of heat

Insulation: Suspended concrete ground floor slabs that are formed on or close to the ground can be formed on top of a d.p.m. and a layer of insulation, providing there is no likelihood of either being damaged by the work of placing the floor slab or of the insulation being saturated with water from the ground.

With a suspended reinforced concrete ground slab the most practical position for the insulation is on top of the slab either underneath a screed as illustrated in Fig. 139, with the d.p.m. joined to the d.p.c. in foundation walls or as part of a floating floor under a boarded finish. Where the d.p.m. is under the insulation the top surface of the concrete slab should be covered with a thin layer of cement and sand levelled off to protect the d.p.m. from damage by irregularities in the top of the slab.

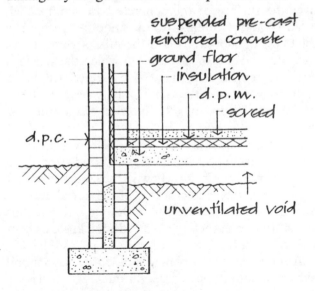

Fig. 139

113

FLOOR FINISHES FOR SOLID CONCRETE FLOORS

For sheds, workshops, stores and garages, the finished top surface of the oversite concrete is sometimes used as the floor surface to save the cost of an applied floor surface, or finish. Concrete is not satisfactory as a floor surface because even though it can be given a smooth finish with a power float, many of the fine particles of sand and cement are brought to the surface. These particles have poor resistance to wear and in a short time the surface of the concrete 'dusts' and requires frequent vigorous brushing. Being a coarse grained material, concrete cannot be washed clean and if it becomes stained the stains are permanent.

A concrete floor finished by power floating is generally a satisfactory base for the thicker floor finishes such as mastic asphalt, thick tiles and wood blocks. For the thin finishes such as plastic, linoleum and rubber sheet and tile, the more precisely level, smooth surface of a screeded base is necessary.

Floor screeds

The purpose of a floor screed is to provide a smooth level surface on which a floor finish can be applied. The usual materials for a floor screed are cement, sand and water which are thoroughly mixed, spread over the surface of the concrete base, compacted, levelled and trowelled to a smooth finish. The thickness of the screed and the mix of cement and sand depends on the surface on which the screed is laid. The cement-rich mix used in a screed will shrink as it dries out and the thinner the screed the more rapidly it will dry and the more it will shrink and crack.

A screed laid on a concrete base, within three hours of placing the concrete, will bond strongly to the concrete and dry slowly with the concrete so that drying shrinkage and cracking of the screed, relative to that of the concrete will be minimised. For this monolithic construction of screed a thickness of 12 of screed will suffice.

A screed laid on a concrete base that has set and hardened should be at least 40 thick. To provide a good bond between the screed and the concrete, the surface of the concrete should be hacked by mechanical means, cleaned and dampened and then covered by a thin grout of water and cement before the screed

is laid. With a good bond to the concrete base a separate screed at least 40 thick will dry sufficiently slowly to avoid serious shrinkage cracking.

Where a screed is laid on an impermeable damp-proof membrane or building paper over an insulating layer, there will be no bond between the screed and the concrete base so that drying shrinkage of the screed is unrestrained. So that the screed does not dry too rapidly and suffer shrinkage cracking, the screed in this unbonded construction should be at least 50 thick.

A screed laid on a layer of compressible thermal or sound insulating material should be at least 65 thick for domestic and 75 for other buildings, if this floating construction is not to crack due to drying shrinkage and the deflection under loads on the floor.

For screeds up to 40 thick, a mix of Portland cement and clean sand in the proportions by weight of 1:3 to $1:4\frac{1}{2}$ is used. The lower the proportion of cement to sand the less the drying shrinkage. For screeds over 40 thick a mix of fine concrete is often used in the proportions of $1:1\frac{1}{2}:3$ of cement, fine aggregate and coarse aggregate with a maximum of 10 for the coarse aggregate.

Screeds should be mixed with just sufficient water for workability. The material is spread over the surface of the base and thoroughly compacted by tamping to the required thickness and level and then is finished with a wood or steel float. A wood float finish is used for wood block and thick tile floors and a steel finish for the thin sheet and tile finishes. The screed should be cured, that is allowed to dry out slowly over the course of several days, by covering it with sheeting, such as polythene, to minimise rapid drying shrinkage and cracking.

There are on the market several screeds designed specifically for use in public places and factories where resistance to wear, freedom from dust, resistance to oils and acids and ease of cleaning are considerations. These screeds of cement rubber-latex, cement bitumen emulsion, cement resin, and resins and binders are spread to a thickness of from 13 to 2 and finished by a trowel on a prepared cement and sand screed for the thin finishes and on a floated concrete base for the thicker materials.

A good floor surface should be smooth, level and sufficiently hard so that it is not worn away for many years. It should not readily absorb water or other liquids likely to be spilled on it and it should be capable of being easily cleaned.

The floor finishes in common use may be classified as:

Jointless
Thin tile and sheet
Wood and wood based
Thick tile

Jointless floor finishes

Granolithic paving: In factories, stores, garages and like buildings, where the floor has to withstand heavy wear, granolithic paving is often used. This consists of a mixture of crushed granite which has been carefully sieved so that the particles are graded from coarse to very fine in such proportions that the material when mixed will be particularly free of voids or small spaces, and when mixed with cement will be a dense mass. The usual proportions of the mix are $2\frac{1}{2}$ of granite chippings to one of Portland cement by volume. These materials are mixed with water and the wet mix is spread uniformly and trowelled to a smooth flat surface. When this paving has dried and hardened it is very hard wearing.

Every material which has a matrix (binding agent) of cement shrinks with quite considerable force as it dries out and hardens. Granolithic paving is rich in cement and when it is spread over a concrete base it shrinks as it dries out and hardens. This shrinkage is resisted by the concrete. If the concrete is dense and hard, and its surface has been thoroughly brushed to remove all dust or loose particles, it will successfully restrain the shrinkage in the granolithic paving. If, however, the concrete on which the granolithic paving is spread is of poor quality, or if the surface of the concrete is covered with dust and loose particles, the shrinkage of the granolithic paving will be unrestrained and it will crack and, in time, break up. On good clean concrete which has dried out the granolithic paving is spread to a thickness of about 20 and is trowelled to a smooth surface.

If the granolithic paving is laid as soon as the oversite concrete is hard enough to stand on, then the paving can be spread 15 thick. This at once economises in the use of the granolithic material, which is expensive, and at the same time, because the wet granolithic binds firmly to the still damp concrete, there is very little likelihood of the paving cracking as it dries out. Granolithic paving which is to be spread over old, poor quality concrete, liable to crumble and dust, must be laid at least 65 thick to prevent it cracking seriously as it dries out.

The reason for this is that if the granolithic paving is spread thinly it will dry out very quickly and shrink fiercely. The shrinkage will not be restrained by the poor concrete below and the paving will crack and fall to pieces. If, however, the paving is 65 thick, or more, it will dry out more slowly and the shrinkage of the top surface, which dries out first, will be restrained by the wetter material below, which takes longer to dry out, and in this way cracking of the material is avoided.

Because the floor layer (pavior) runs screeds to obtain a true level finish this paving is sometimes described as granolithic screed, or a screed of granolithic. As has been said, the finished surface of granolithic paving is smooth, dense and hard wearing but with very heavy wear the material will break up very slightly in time and form dust. This dust may cause serious damage to some machinery and make working conditions unpleasant. There are on the market a number of preparations which, if added to the granolithic paving material, will considerably reduce the amount of dust caused by wear on the surface of the paving. These proprietary preparations vary considerably in composition and generally have the effect of sealing the minute pores in the material, or of forming a hard, water-repellent skin on the surface. These preparations are variously described as 'sealers' or 'hardeners'. The resistance to wear on the surface of granolithic paving can be improved by sprinkling the surface, whilst it is still wet, with fine carborundum. The carborundum is then trowelled in. Carborundum is very hard and because it is expensive it is applied only as a very thin top dressing.

Mastic asphalt flooring: Mastic asphalt for flooring is made from either limestone aggregate, natural rock or black pitch-mastic in the natural colour of the material or coloured. Mastic asphalt serves both as a floor finish and a damp-proof membrane. It is a smooth, hardwearing, dust free finish, easy to clean but liable to be slippery when wet. The light duty grade is fairly readily indented by furniture. Mastic asphalt has been less used as a floor finish since the advent of the thin plastic tiles and sheets.

The light duty, non-industrial grade, which is laid in one coat to a finished thickness of from 15 to 20, is used for offices, schools and housing. The medium grade is laid in one coat to a thickness of 20 to 25 and the heavy duty in one coat to a thickness of 30 to 50. Mastic asphalt can be laid on a level, power floated concrete finish or on a level, smooth cement and sand screed. The asphalt finish can be coloured in one of the red or brown shades available.

Magnesite floor finish: The finish is composed of burned magnesite, fine sawdust, wood flour and water to which is added some magnesium chloride solution. The proportions of the materials are 9 magnesite, 2 sawdust, 1 wood flour. The magnesium chloride combines with the magnesite to form magnesium oxychloride which hardens and acts as a very hard, strong cement.

The materials are mixed on the site, preferably by machine, and the magnesium chloride is added and mixed in. This forms a mass of putty-like consistency which is spread evenly over the surface to be covered, and is trowelled flat and level. The material gradually hardens and when it is hard enough to stand the weight of a board, with a man on it, the surface is scraped to remove trowel marks and then trowelled again. The surface hardens in about forty-eight hours and is then sealed with a chloride solution. This finish can be laid in one or two coats. One coat is laid to a finished thickness of about 20 and for the two coat work a 15 thick undercoat and 5 finishing coat are laid. The two coat work is used when a grained or mottled surface is required.

The finish can be pigmented to give many different colours among which red and buff are popular. The finished surface can be mottled and grained with various colours in imitation of marble. Magnesite finish can be laid on a clean level concrete floor or on a screed of cement and sand. This floor finish is hard and resistant to normal wear on domestic and office floors. It does not crack or powder, if properly laid, and being resistant to water, mineral oils and fats can be washed and if oiled is easily kept clean.

Magnesite floor finish can only be successfully laid by craftsmen trained in the use of the materials. Of late years this finish has lost popularity because it has been given a bad name due to serious failures of the surface finishes when laid by men not competent in the use of the materials.

Thin tile and sheet materials

Plastic floor finishes: Thin tiles and sheets of plastic are much used as floor finishes for houses, offices, schools and public buildings because of low initial cost and ease of cleaning. The materials available are:

Thermoplastic tiles
PVC (vinyl) asbestos tiles
Flexible PVC tiles and sheets
Flexible PVC sheets on a backing of felt or cellular PVC.

Thermoplastic tiles are made from a blend of mineral asphalt and/or thermoplastic resins, such as those obtained from the distillation of coal or oil, fibres, fillers and pigments. The standard size of tile is 300 and 250 square in thicknesses of 2.5 mm and 3.0 mm. The standard tile is used in offices and schools where grease and oil, which may cause deterioration, are not present. The tiles are not as flexible for laying as other plastic materials and are somewhat hard and noisy underfoot.

Thermoplastic tiles – PVC modified, have better resistance to grease and oil and are slightly more flexible than the standard grade.

PVC (vinyl) tiles, sometimes termed vinyl tiles or asbestos tiles, are made from a blend of thermoplastic binder (either vinyl chloride polymer or vinyl chloride copolymer or both), fibres, fillers and pigment in sizes of 225 and 300 square and thicknesses of 1.6, 2.0, 2.5, 3.0 and 3.2 mm. These tiles have better resistance to grease and oil than thermoplastic tiles, have only moderate flexibility and are somewhat hard and noisy underfoot.

Flexible PVC tiles and sheets are made from a blend of thermoplastic binder (either vinyl chloride polymer or vinyl chloride copolymer or both), fillers and pigments in tiles 225, 250 and 300 square and sheets 1.2, 1.5, 1.8, 2.0 and 2.1 m wide and thicknesses of 1.5, 2.0, 2.5 and 3.0 mm. Because of their flexibility these tiles and sheets are easier to lay than PVC asbestos tiles and having a brighter clearer colour, are often preferred.

Flexible PVC sheets with a backing are made from a blend of polymeric materials (either vinyl chloride polymer or vinyl chloride copolymer or both), with fillers and pigments and a backing of felt or cellular PVC. The felt backing is of needle-loom felt and the sheets are 1.5 and 2.0 m wide. The foam backing is of plasticised polymeric material, stabilised and converted to a cellular state and the sheets are 1.2, 1.5, 1.8, 2.0 and 2.1 m wide and 2.0, 2.5, 3.0 and 4.5 mm thick. These flexible sheets can be heat welded together to make a seamless floor. This floor is hardwearing and soft and quiet underfoot due to the backing and is much used in houses, hospitals and public buildings for ease of cleaning and sound deadening effect.

Thermoplastic and PVC tiles and sheets should be laid on a carefully prepared smooth surface screed, as even small variations in the surface of the screed will show through these thin materials. The tiles and sheets are bonded with an adhesive that is spread evenly, by trowel, over the screed and the tiles and sheets are then pressed into place. These tiles and sheets should be warmed before laying in cold conditions. The flexible backed PVC sheets are laid on a carefully prepared screed and bonded to an adhesive that is spread evenly by trowel and the sheets are then rolled to make a good bond to the adhesive.

Plastic floor finishes do not dust, are easily cleaned and may be polished for appearance sake and ease of cleaning. They tend to be slippery when wet, wear moderately well and quickly show scuff marks and the indent of heavy furniture and scratches.

Linoleum is made from powdered cork, fillers and pigments and oxidised linseed oil with a backing and reinforcement of jute canvas supplied in sheet and tile form in widths of 1830 for sheets and 300 square tiles and thicknesses of 2.0, 2.5, 3.2 and 4.5 mm. This resilient, hard wearing, easy to clean floor finish is less used than it was, having been replaced largely by plastic sheets and tiles. It requires more skill in cutting and laying than plastic flooring. It does not dust, can be cleaned by washing and is fairly readily indented and scratched. With reasonable care it is an admirable, durable floor finish.

Wood and wood-based finishes

Wood floor finishes for concrete floors: Wood as a floor finish is prepared as either strip flooring or as wood blocks. Generally wood strip flooring, which is expensive, is used where wear on the surface is light, as in houses, and wood blocks for houses, public buildings, offices and schools.

Wood strip flooring: Strips of hardwood or softwood of good quality, specially selected so as to be particularly free of knots, are prepared in widths of 90 or less and 19, 21 or 28 in thickness. The type of wood chosen is one which is thought to have an

attractive natural colour and decorative grain. The edges of the strip are cut so that one edge is grooved and the other edge tongued, so that when they are put together the tongue on one fits tightly into the groove in its neighbour as in Fig. 140. The strips are said to be tongued and grooved, usually abbreviated to T & G. The main purpose of the tongue and groove is to cause the strips to interlock so that any slight twisting of one strip is resisted by its neighbour.

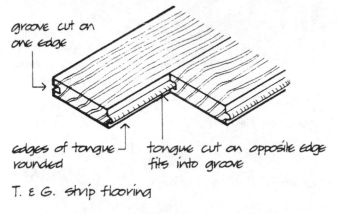

T. & G. strip flooring

Fig. 140

There is always some tendency for wood strips to twist out of flat, due to the wood drying out and to resist this the strips have to be securely nailed to wood battens which are secured to the concrete floor either by means of galvanised metal floor clips, or in a cement and sand screed. The drawing of part of a concrete floor finished with wood strips nailed to battens will explain the arrangement of the parts, Fig. 141. The floor clips are of galvanised sheet steel which is cut and stamped to the shape shown in Fig. 141. These are usually set into the concrete floor whilst it is still wet. They are placed in rows 450 apart and the clips in each row are spaced 750 apart so that when the concrete has hardened, the clips are firmly bedded in it. The strip flooring is usually laid towards the end of building operations and the 50 × 38 or 50 × 25 softwood battens are wedged up until they are level and the clips are then nailed to them. The strip flooring is then nailed to the battens. An alternative method of securing the battens to the concrete sub-floor is to bed dovetailed battens in a sand and cement screed spread on the concrete as illustrated in Fig. 142.

If timber is in contact with a damp surface it will rot and it is important to protect both the battens and the strip flooring from damp which may rise from or through the concrete sub-floor. The battens should

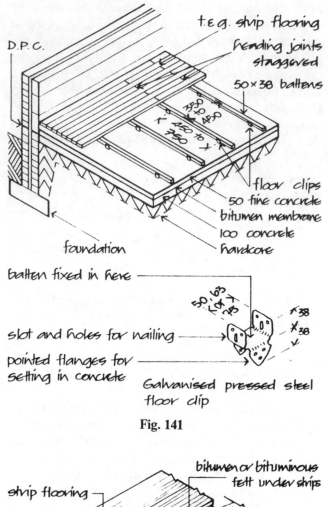

Fig. 141

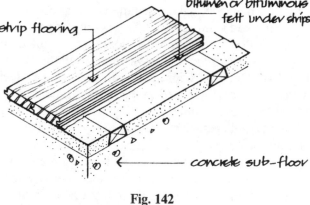

Fig. 142

through the tongues into the battens below so that the groove in the edge of the next board hides the nail.

Even though the nails used have small heads they may split the narrow tongue off the edge of the strip and this makes a poor fixing. To avoid this, the edges of the strips can be cut with splayed tongued and grooved joints. The finished surface of the wood strips is usually wax polished to protect it, show the grain and colour of the wood to its best advantage and to make it easy to clean the floor. Oak, beech, maple and birch are commonly used for strip flooring.

Wood block floor finish: Blocks of some wood with good resistance to wear are cut. The blocks are usually 229 to 305 long by 75 wide by from 21 to 40 thick. The blocks are laid on the floor in bonded, herringbone or basket weave pattern.The usual patterns are illustrated in Fig. 143.

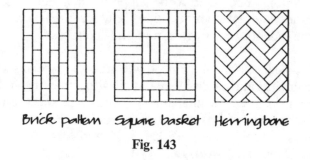

Fig. 143

Wood blocks are laid on a thoroughly dry, clean, level cement and sand screeded surface which has been finished with a wood float to leave its surface rough textured. A thin layer of bitumen is spread over the surface of the screed and into this the blocks are pressed. The lower edges of the blocks of wood are usually cut with a half dovetail incision so that when the blocks are pressed into the bitumen some bitumen squeezes up and fills these dovetail cuts and so assists in binding the blocks to the bitumen, as shown in Fig. 144.

If the wood blocks have been thoroughly seasoned (dried) and they are firmly pressed into the bitumen they will usually be securely fixed to the floor. It is possible, however, that one or more blocks may not be firmly fixed and will come up. To prevent this happening good quality wood blocks 25 thick and over have either tongues and grooves cut on their edges or wood dowels to joint them as shown in Fig. 144. When the blocks have been laid the surface is usually 'sanded'. The word 'sanded' describes the

be impregnated with a preservative before they are fixed and either the surface of the concrete sub-floor should be covered with a coat of bitumen or a waterproof membrane should be used in or under the concrete oversite. As wood strip flooring is an expensive, decorative, floor finish the strips of wood are nailed to the battens so that the heads of the nails do not show on the finished surface of the floor. This is termed secret nailings. If the strips have tongued and grooved edges the nails are driven obliquely

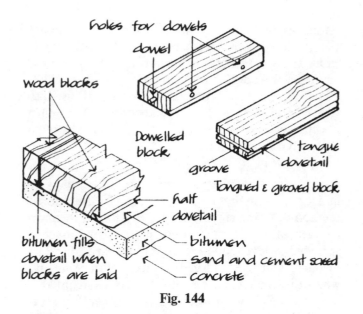

holes for dowels
dowel
wood blocks
Dowelled block
tongue
dovetail
groove
Tongued & grooved block
half dovetail
bitumen fills dovetail when blocks are laid
bitumen
sand and cement screed
concrete

Fig. 144

operation of running a power driven sanding machine over the surface. The machine has a drum surfaced with sand paper, which sandpapers the surface until it is smooth. The surface is then either oiled or wax polished.

Chipboard is the common name for particle boards that are made from particles of wood bonded with a synthetic resin, usually urea formaldehyde, and compressed into rigid boards.

The various densities of chipboard that are manufactured are differentiated as:

Type I	Standard for general purposes
Type II	Flooring
Type III	Improved moisture rating
Type II/III	Combines strength of II with moisture resistance of III

These cheap, rigid, stable boards have a reasonably smooth surface, can be cut, nailed and fixed like wood and are much used as a floor finish or floor deck for applied finishes such as thin tiles, sheets and carpet. On timber floors chipboards 18 and 22 thick are used on joists at 400 and 600 centres with boards 2400 long by 1200 wide.

On concrete floors chipboards are fixed to battens in floor clips or battens nailed to the concrete base in the same way that boards are fixed. The advantage of the chipboard floor is that electrical, heating and water services can be accommodated between the battens and the boards. For use as a floor finish, chipboard can be sealed to prevent dusting, to facilitate cleaning and to avoid staining.

Thick tiled finishes

Tiled floor finishes: The word tile was first used to describe thin slabs of burned clay, usually rectangular in shape, which were used to cover roofs and floors. Similarly shaped slabs of other materials are today described as being tiles, for example, concrete tiles and rubber tiles.

Clay floor tiles: The two different types which are manufactured today are distinguished as floor quarries and clay floor tiles. The word quarry is derived from the French *carre*, meaning square. Clay floor quarries are thicker, less uniform in shape and size and have a less smooth surface than clay floor tiles.

Floor quarries are manufactured in Staffordshire and Wales from natural plastic clays. The clay is ground and mixed with water and then moulded in hand operated presses. The moulded clay tile is then burned in a kiln. If the clay is of good quality and the tile is burned at the correct temperature the finished tile will be very hard, dense and will wear extremely well. But as there is no precise examination of the clays used, nor accurate control of pressing or burning, the tiles produced vary considerably in quality, from very hard well burned quarries, to soft underburned quarries unsuitable for any use in buildings. The manufacturers grade the tiles according to their hardness, shape and colour. The first or best quality of these clay floor quarries are so hard and dense that they will suffer the hardest wear on floors for centuries without noticeably wearing. Because they are made from plastic clay, which readily absorbs moisture, quarries shrink appreciably when burned and there is often a noticeable difference in the size of individual tiles in any batch. The usual colours are red, black, buff and heather brown and the tiles are manufactured in the following sizes:

Preferred modular co-ordinating sizes
  200 × 200
  200 × 100
  100 × 100
Non-modular sizes
  229 × 110 × 29
  194 × 194 × 25
  194 × 194 × 12.5 mm
  150 × 150 × 12.5 mm
  150 × 72 × 12.5 mm
   94 × 94 × 12.5 mm

Clay floor tiles: Three types of these tiles are manufactured, plain colours, vitreous and encaustic.

Plain colours are manufactured from natural clays selected for their purity. The clay is ground to fine dry powder and a small amount of water is added. The damp powder is heavily dust pressed into tile shape and the moulded tiles are burned. Because finely ground clay is used the finished tiles are very uniform in quality and because little water is used in the moulding, very little shrinkage occurs during burning. The finished tiles are uniform in shape and size and have smooth faces. The tiles are manufactured in red, buff, black, chocolate and fawn and in the following sizes:

Preferred modular co-ordinating sizes
200 × 200
200 × 100
100 × 100

Non-modular sizes
300 × 300 × 8
152 × 152 × 12.5 mm
152 × 152 × 9.5 mm
150 × 150 × 12
150 × 150 × 9
100 × 100 × 9

Vitreous floor tiles: The word vitreous means glass-like. These tiles are made from felspar which melts when the tile is burned and causes it to have a hard, smooth, glass-like surface which is impervious to water. By itself felspar would make the tile too brittle for use and it is mixed with both clay and flint. These materials are ground to a fine powder, a little water is added and the material is heavily pressed into tile shape and then burned.

The tiles are uniform in shape and size and have a very smooth semi-gloss surface which does not absorb water or other liquids. Colours and sizes are the same as those of plain colour tiles. Vitreous floor tiles are more expensive than plain colour floor tiles. These tiles are called tessellated by some manufacturers. The word tessellated derives from tesserae, meaning small pieces of tile. Other manufacturers call these tiles ceramic which means 'made from clay'. It is best to avoid the use of these misleading names.

Encaustic tiles: The word 'encaustic' is used to describe a tile which has a pattern, or design, inlaid in its surface in differently coloured clays. An ordinary plain colour tile is first press moulded with one or more shallow sinkings in its surface and into these sinkings are pressed one or more coloured clays. Clays of many different colours can be used to decorate the surface of the tiles. the tiles are burned in a kiln. The sizes of these tiles are the same as plain colour tiles.

Laying clay floor tiles: Clay floor tiles should be laid on a solid base of concrete. The traditional method of laying clay floor tiles was to bed them on a layer of wet cement and sand spread over a screeded or level concrete floor. This **direct bedding method** is satisfactory for small areas of flooring and where the concrete base is fully dried out and unlikely to suffer shrinkage movement. Where relative movement is likely, the direct bedding method may well be unsatisfactory. To accommodate relative movements between the concrete base, the screed and the tiles, it is good practice to bed the tiles over a separating layer on a thick bed or on a thin bed adhesive. The relative movements that can occur are drying shrinkage of concrete and screed bases, differing thermal movements and creep of concrete. Where clay tiles are firmly bonded to the concrete or screed there is a likelihood of differential movement between the base and the tiles causing tiles to arch or ridge upwards so causing failure of the floor. Tiles should, therefore, be bedded in such a way as to accommodate possible movement.

Bedding on a separating layer: A separating layer of sheet material such as bituminous felt, polythene or building paper is spread over the power floated concrete or screeded base with the sheets lapped to prevent any adhesion of the bedding to the base. The tiles are then bedded on a layer of cement and sand about 25 thick on which they are levelled and the joints grouted or filled, as illustrated in Fig. 145.

Thick bed semi-dry mix method: A bed of semi-dry cement and sand, mix 1:3½ or 1:4, is spread over the concrete or screed base and packed to a thickness of at least 40 . The bed is then covered with a grout (wet mix) of cement and sand, mix 1:1 or a grout of neat cement and water into which the tiles are bedded, levelled and the joints grouted or filled. The semi-dry bed accommodates relative movements between the base and the tiles.

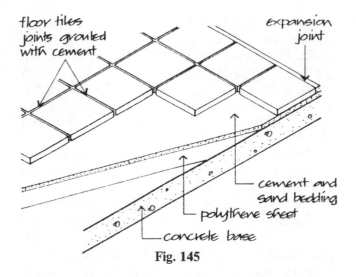

floor tiles
joints grouted
with cement

expansion
joint

cement and
sand bedding
polythene sheet
concrete base

Fig. 145

Thin bed adhesive fixing: Tiles are bedded on a thin bed of a proprietary cement-based adhesive or epoxy resin adhesive spread to a thickness of between 3 and 6. The adhesive provides a firm resilient base for the tiles that will accommodate movement.

Movement joints: To allow for possible relative movements, a movement joint should be formed around the perimeter of a tiled floor as illustrated in Fig. 145. The joint is filled with an elastic sealing compound. For large areas of tiled floor, additional expansion joints should be formed both along and across the floor at intervals.

Joints between tiles: The joint between tiles should be wide enough to allow for variations in the size of tiles and a joint of at least 3 for small tiles and up to 15 for larger tiles is necessary. Wide joints may be filled with a grout of cement or a mix of cement and fine sand or one of the cement-based or acrylic or epoxy resin grouts for narrow joints.

Concrete tiles: These tiles are manufactured from clean sand and cement with a surface dressing of coloured cement and ground flint. The materials are mixed wet and heavily moulded into tile shape and allowed to dry out and harden slowly. The finished tiles are very hard and dense and are finished in a variety of colours. Sizes of tiles are $300 \times 300 \times 25$, $225 \times 225 \times 19$ and $150 \times 150 \times 16$. The tiles are laid on a bed of cement and sand over a separating layer of polythene or building paper on a concrete or screeded surface. These tiles were used as a cheap substitute for clay tiles before the advent of thermoplastic tiles.

SUSPENDED TIMBER GROUND FLOORS

Many houses built in this country from about 1820 up to about 1939 were constructed with timber ground floors raised 300 or more above the site concrete or packed earth or brick rubble below. The purpose of raising the ground floor was to have the floor surface of the ground floor living rooms sufficiently above ground level to prevent them being cold and damp in winter. At that time imported softwood timber was cheap and this ground floor construction was both economical and satisfactory.

Since the end of the Second World War (1945), imported softwood timber has been expensive and for some years after the war its use was restricted by government regulations. Most ground floors today are formed directly off the site concrete and this is covered with one of the surface finishes as described previously.

Strength and stability

A suspended or raised timber ground floor is constructed as a timber platform of boards nailed across timber joists bearing on $\frac{1}{2}B$ brick walls raised directly off the site concrete as illustrated in Fig. 146. The raised timber floor is formed inside the external walls and internal brickwork partitions, and is supported on brick sleeper walls.

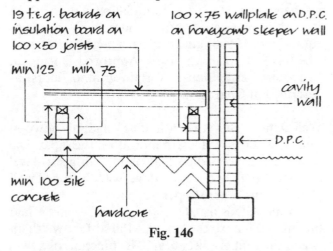

19 t.eg. boards on insulation board on 100 × 50 joists

min 125   min 75

100 × 75 wallplate on D.P.C. on honeycomb sleeper wall

cavity wall

D.P.C.

min 100 site concrete

hardcore

Fig. 146

Sleeper walls are $\frac{1}{2}B$ thick and are built directly off the site concrete up to 1.8 apart. These sleeper walls are generally built at least three courses of bricks high and sometimes as much as 600 high. The walls are built honeycombed to allow free circulation of air below the floor, the holes in the wall being $\frac{1}{2}B$ wide by 85 deep as shown in Fig. 147.

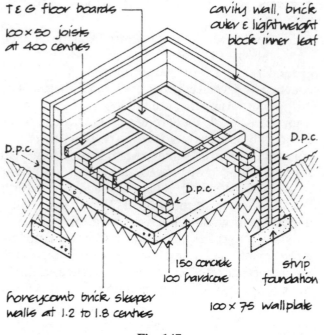

T & G floor boards

100 × 50 joists at 400 centres

cavity wall, brick outer & lightweight block inner leaf

D.p.c.

D.p.c.

D.p.c.

150 concrete
100 hardcore

strip foundation

honeycomb brick sleeper walls at 1.2 to 1.8 centres

100 × 75 wallplate

Fig. 147

The practical guidance in Approved Document C to The Building Regulations 1985 requires a space of at least 75 from the top of the concrete to the underside of a wall plate and at least 125 to the underside of the floor joists, concrete oversite at least 100 thick on a hardcore bed or concrete 50 thick on a d.p.m. of 1000 gauge polythene on a bed which will not damage it. Underfloor ventilation should have a free path between opposite sides, with openings equivalent to 3000 square for each metre run of wall. The Revised document 1989, Approved Document L, requires insulation to certain ground floors, as described in Chapter 1.

Wall plate: This is a continuous length of softwood timber which is bedded in mortar on the d.p.c. The wall plate is bedded so that its top surface is level along its length and it is also level with the top of wall plates on the other sleeper walls.

A wall plate is usually 100 × 75 timber and is laid on one 100 face so that there is 100 surface width on which the timber joists bear. The function of a wall plate for timber joists is two-fold. It forms a firm level surface on which the timber joists can bear and to which they can be nailed and it spreads the point load from joists uniformly along the length of the wall below.

Floor joists: These are rectangular section soft-

wood timbers laid with their long sectional axis vertical and laid parallel spaced from 400 to 600 apart. Floor joists are from 38 to 75 thick and from 75 to 225 deep. The span of a joist is the distance measured along its length between walls that support it. The sleeper walls built to support the joists are usually 1.8 apart or less, and the span of the joists in this type of floor is therefore 1.8 or less.

Timber, chipboard or plywood boards are laid across the joists and they are nailed to them to form a firm level floor surface.

From a calculation of the dead and imposed loads on the floor the most economical size and spacing of joists can be selected from the tables in Approved Document A to The Building Regulations 1985 and from this the spacing of the sleeper walls to support the joists.

Similarly the thickness of the floor boards to be used will determine the spacing of the joists, the thicker the board the greater the spacing of the joists.

Floor boards: Any length of timber 100 or more wide and under 50 thick is called a board. Floor boards for timber floors are usually 16, 19, 21 or 28 thick and 65, 90, 113 or 137 wide and in length up to about 5.0. The boards are cut from whitewood which is moderately cheap or from redwood which is more expensive but which wears better. The edges of the boards may be cut square, or plain edged, which is the cheapest way of cutting them. But as the boards shrink ugly cracks may appear between them.

The most usual way of cutting the boards is with a projecting tongue on one edge and a groove on the opposite edge of each board, as in Fig. 140. The boards are then said to be tongued and grooved abbreviated to T & G. The boards are laid across the floor joists and cramped together. Cramping describes the operation of forcing the edges of the boards tightly together so that the tongues fit firmly into the grooves and there are no open cracks between the boards. The boards as they are cramped up are nailed to the joists with two nails to each board bearing on each joist.

Heading joints: Floors of small rooms can often be covered with boards sufficiently long to run in one length from wall to wall but in most rooms the ends of boards have to be cut to butt together. The joint between the end of one board and the end of another is described as the heading joint. The appearance of a boarded floor is spoiled if the heading joints run in

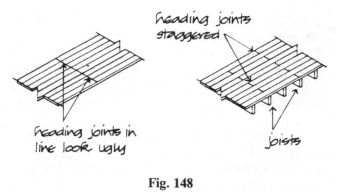

Fig. 148

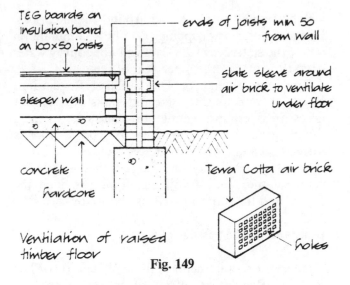

Ventilation of raised timber floor

Fig. 149

a continuous line across the floor because the cut ends of the boards tend to be somewhat ragged and the continuous joint looks ragged and ugly (Fig. 148). The heading joints in floor boards should always be staggered in some regular manner. Obviously the heading joint ends of boards must be cut so that the ends of both boards rest on a joist to which the ends are nailed. A usual method of staggering heading joints is shown in Fig. 148.

End matched flooring: Some hardwood strips are prepared with tongues and grooves on the ends of the strips so that their ends firmly interlock and do not have to lie over a joist. The strips of flooring are said to be end matched. The end joint so formed is much neater than a sawn end of ordinary boarding.

End support of floor joists: The floor joists of the raised timber ground floor bear on wall plates on sleeper walls and the best method of supporting the ends of the joists at external walls and at internal brick partitions is to build a honeycombed sleeper wall some 50 away from the wall to carry the ends of the joists, as shown in Fig. 149. The sleeper wall is built away from the main wall to allow air to circulate through the holes in the honeycomb of the sleeper wall. The ends of the joists are cut so that they are clear of the inside face of the wall by 50.

Resistance to ground moisture

Damp-proof course

D.p.c.: A d.p.c. should be spread and bedded on top of the sleeper walls to prevent any moisture rising through site concrete without a d.p.m. and through sleeper walls to the timber floor. Any of the materials described in Chapter 1 may be used for this purpose.

Ventilation of space below raised timber ground floor: The space below this type of floor is usually ventilated by forming ventilation gratings in the external walls below the floor so that air from outside the building can circulate at all times under the floor. The usual practice is to build air bricks into the external walls. An air brick is a special brick made of terra cotta (meaning earth burned) with several square or round holes in it and its size is 215×65, 215×140 or 215×215 (Fig. 149). These bricks are built into external and internal walls for each floor to provide 3000 square of ventilation for each metre run of wall. The bricks are built in just above ground level and below the floor as in Fig. 149.

The purpose of the air bricks is to cause air to circulate under the floor and so avoid stagnant damp air which is likely to induce the dry rot fungus to grow. The disadvantage of ventilating the space below this type of floor is that in winter the floor is liable to be cold. It is usual, therefore, to fix an insulating board or quilt under the floor boards to minimise transfer of heat to the ventilated space below the floor.

Air bricks in cavity walls: It is common practice today to prevent cold air from outside entering the cavity of a cavity wall to avoid the inner skin becoming cold. When ventilating air bricks are built into a cavity wall some means must be devised of preventing cold air getting into the cavity through the air brick. One common method of doing this is to build a roofing slate sleeve around the air bricks and across the cavity or a short length of pipe as shown in Fig. 149. A duct is made out of four pieces of slate which are built into the wall around the air bricks. Providing mortar droppings do not accumulate on top of the slate sleeve the slate will not convey water from the outer to the inner skin of the wall.

The requirements of The Building Regulations 1985 for resistance to ground moisture do not necessarily require ventilation below suspended timber ground floors. Where this type of floor is formed on a layer of oversite concrete and a d.p.m. adequate to resist moisture, there is then no need to ventilate the space below the floor provided the d.p.m. and the d.p.c. in walls is continuous.

Resistance to the passage of heat

To meet the requirements of the 1989 revision of The Building Regulations 1985 for resistance to heat transfer through ground floors, it is necessary to insulate suspended timber ground floors that have a ventilated space below. The most practical way of insulating a suspended timber ground floor is to fix mineral wool roll, mat or quilt or semi-rigid slabs between the joists. Rolls or quilt of loosely felted glass fibre or rockwool are supported by a mesh of plastic that is draped over the joists and stapled in position to support the insulation. Semi-rigid slabs or batts of fibre glass or rockwool are supported between the joists by nails or battens of wood nailed to the sides of the joists.

UPPER FLOORS

Timber floors

Timber upper floors for houses and flats are about half the cost of comparable reinforced concrete floors. For upper floors of offices, factories and public buildings timber floors are not much used today because the resistance to fire of a timber floor, plastered on the underside, is not sufficient to comply with building regulations for all but small buildings. Concrete floors are used instead because of their better resistance to fire, and better resistance to sound transmission.

Strength and stability

Floor joists: The floor is framed, or carcassed, with softwood timber joists which are usually 38 to 75 thick and 75 to 225 deep. The required depth of joists depends on the dead and imposed loads and the

span. The spacing of the joists is usually from 400 to 600 measured from the centre of one joist to the next. Tables in Approved Document A to The Building Regulations 1985 set out the required size of timber joists for given spans and with given spacing of joists for various loads for single family dwellings of up to 3 storeys.

To economise in the use of timber the floor joists of upper floors usually span (are laid across) the least width of rooms from external walls to internal load bearing partitions. The joists in each room span the least width. The maximum economical span for timber joists is between 3.6 and 4.0. For greater spans than 4.0 it is economic to reduce the span of the joists by the use of steel beams.

Double floors: Where the span of a timber floor is greater than the commercially available length of timber and where, for example, joists span parallel to a cross wall, it is convenient and economic to use a steel beam or timber beam to support timber joists. This combination of a beam and the joists is described as a double floor. Steel beams are generally used because of their smaller section. The supporting steel beam may be fixed under the joists or wholly or partly hidden in the depth of the floor. To provide a fixing for the ends of the joists, timber plates are bolted to the flange of the beam and the joists fixed to the plates as illustrated in Fig. 150.

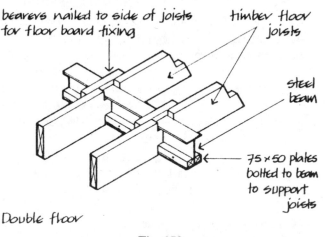

Double floor

Fig. 150

Strutting between joists: When timber is seasoned it shrinks, and timber such as floor joists, which is not cut on the radius of the circle of the log, does not shrink uniformly. The shrinkage will tend to make the floor joists twist, or wind, and to prevent cracking of a plaster ceiling, which this twisting would cause,

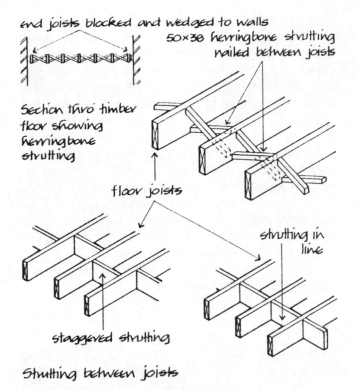

end joists blocked and wedged to walls

50×38 herringbone strutting nailed between joists

Section thro' timber floor showing herringbone strutting

floor joists

strutting in line

staggered strutting

Strutting between joists

Fig. 151

timber strutting is used. The type most commonly used is that known as herringbone strutting. This consists of short lengths of softwood timber about 50 × 38 nailed between the joists as in Fig. 151. The strutting is fixed between all the joists right across each room from wall to wall. Wedges and blocks of wood are fixed between the end joists and the wall as in Fig. 151, to tighten the system of strutting. Alternatively a system of solid strutting is sometimes used.

This consists of short lengths of timber of the same section as the joist which are nailed between the joists either in line or staggered, as in Fig. 151. This is not usually so effective a system of strutting as the herringbone system, because unless the short solid lengths are cut very accurately to fit to the sides of the joists they do not firmly strut between the joists.

As with herringbone strutting the end joists are blocked and wedged up to the surrounding walls.

Usually one set of struts is used for joists spanning up to 3.6, and two for joists spanning more than 3.6. A single set of struts is fixed across the floor at mid span.

End support for floor joists: For stability, the end of floor joists must have adequate support from walls or beams. If the floor is to be durable, timber joists should not be built into external walls where their ends may be persistently damp and suffer decay. Timber joists should not be built into or across separating or compartment walls where they may encourage spread of fire. Floor joists are, therefore, either built into internal and external walls or they are supported on hangers, or corbels projecting from the face of walls.

Timber floor joists that are built into walls should bear on a wall plate of timber or metal, which serves to spread the load from the floor along the length of the walls and as a level bed on which the joists bear.

Timber wall plates are of sawn softwood 100 × 65 to course into brickwork, and laid with one 100 face horizontal. The wall plate is bedded level in mortar to take the ends of the joists which are nailed in position to the timber plate and the wall is then raised between and above the floor as illustrated in Fig. 152, which illustrates joists built into an internal load bearing brick wall.

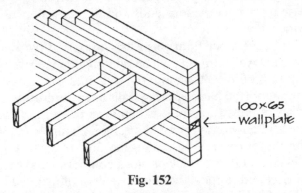

100×65 wall plate

Fig. 152

When timber joists are built into the inner skin of a cavity wall the joists must not project into the cavity and it is wise to treat the ends of the joists with a preservative against the possibility of decay due to moisture penetration. The ends of joists built into a cavity wall may bear on a timber wall plate which should also be treated with a preservative. The wall plate is bedded on the blockwork inner skin. As an alternative, a mild steel bar 75 × 6 may be used. This metal wall plate is tarred and sanded and bedded level in mortar, and the joist ends bear on the plate as illustrated in Fig. 153.

Instead of using a timber or a metal wall plate the joists may bear directly on the brick or block wall with tile or slate slips in mortar packed under each joist end to level the joists. This is a somewhat laborious procedure and the slips may be displaced and the joists move out of level during subsequent building operations. Timber joists may be built into

125

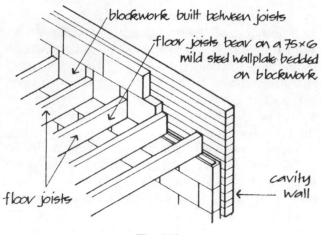

blockwork built between joists

floor joists bear on a 75×6 mild steel wallplate bedded on blockwork

floor joists

cavity wall

Fig. 153

a solid external wall if the wall is thick enough to prevent penetration of moisture to the joist ends and where the wall is protected externally with slate or tile hanging. As a precaution against the possibility of decay due to unforeseen moisture penetration it is wise to protect the joist ends and timber wall plates with a preservative.

Where timber floor joists are supported by solid external walls through which rain might penetrate to the ends of joists built into the wall, and where joists are supported by separating walls as in the cross wall form of construction, it is common to support the joists in galvanised pressed steel joist hangers made for the purpose. These hangers, illustrated in Fig. 154, are built into brick or block courses so that they project and support the ends of joists as illustrated in Fig. 153. Either the hangers are built in as the brick or block wall is raised and the joists fitted later, or the

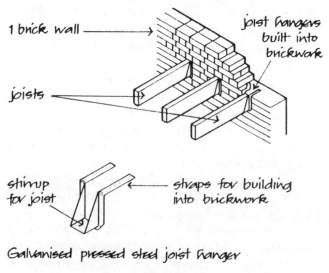

1 brick wall

joist hangers built into brickwork

joists

stirrup for joist

straps for building into brickwork

Galvanised pressed steel joist hanger

Fig. 154

joists, with the hangers nailed to their ends, are given temporary support as the brick or blockwork is built, and the hangers built into horizontal courses to accurately locate them in position. Metal hangers may be used to support joists on the inner skin of a cavity wall providing the span of the floor and the loads on the floor are not such as to be likely to cause so great an eccentric load likely to overturn the wall.

As an alternative to hangers, timber floor joists may be supported by a timber wall plate carried on iron corbels built into walls as illustrated in Fig. 155, or on brick courses corbelled out from the wall as illustrated in Fig. 155. The disadvantage of these two methods of support is that they form a projection below the ceiling. Since the advent of metal joist hangers, corbel bracket and corbel course support for joists have been little used.

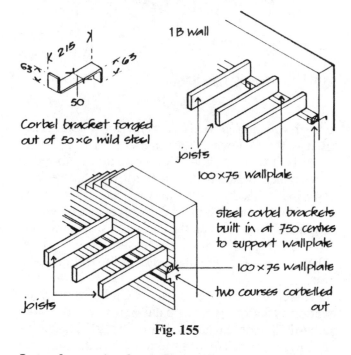

1B wall

215

63   63

50

Corbel bracket forged out of 50×6 mild steel

joists

100×75 wallplate

steel corbel brackets built in at 750 centres to support wallplate

100×75 wallplate

two courses corbelled out

joists

Fig. 155

Lateral restraint for walls: Where a timber floor gives lateral restraint to a wall to stiffen it, the timber joists have to be firmly anchored to the wall to give positive support. Metal ties are screwed or bolted to the ends of joists and the turned down end of the ties built into the wall as illustrated in Fig. 70 or restraint type joist hangers are used as illustrated in Fig. 156. Where a timber floor gives lateral support to a wall parallel to the span of the joists then ties are secured across the joists and the ties anchored to the wall as illustrated in Fig. 70. Similarly, concrete floors giving lateral support are built in with at least 90 bearing at ends.

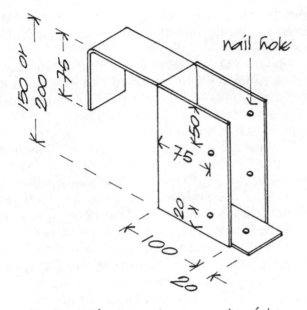

Galvanised steel restraint joist hanger

Fig. 156

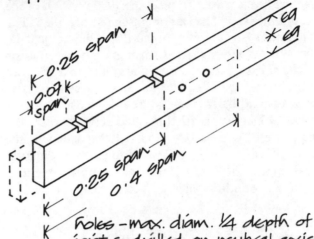

notches – max. depth ⅛th depth, cut no closer than 0.07 to, nor nearer than ¼ span from support

holes – max. diam. ¼ depth of joist & drilled on neutral axis and min. 3 diameters apart centre to centre and between 0.25 and 0.4 span from support

Notches & holes in timber joists

Fig. 157

Notches and holes: So that notches and holes cut in timber joists, do not seriously weaken the strength of the floor, the following limitations are given as practical guidance to meetings the requirements of The Building Regulations 1985 for small domestic buildings.

Notches and holes cut in timber floor joists for water, heating and gas service pipes and for electrical cables should be no deeper than one-eighth of the depth of the joist and not cut closer to a support than 0.07 of the span, nor further away than one-quarter of the span as illustrated in Fig. 157.

Similarly holes cut in joists for pipes or electrical cables should be of no greater diameter than one-quarter of the depth of the joist, should be drilled on the neutral axis and not less than 3 diameters apart measured from centre to centre and located between 0.25 and 0.4 times the span from supports as illustrated in Fig. 157.

Floor boards: The surface of the floor may be formed by nailing boards across the floor joists. As with raised timber ground floors, the boards are usually 19 or 21 thick and have tongued and grooved edges. The boards are cramped up, and nailed and

the heading joints of boards are staggered as described for raised timber ground floors.

Boards of wood, chipboard or plywood are often used as a base for carpet or plastic sheet finishes, instead of narrow boards, for ease of handling and fixing and to minimise joints between boards. Whole sheets of chipboard or plywood are used, with either square edge joints or tongued and grooved joints.

Fire resistance

Structural floors of dwelling houses of two or three storeys are required to have a minimum period of fire resistance of half an hour. Timber floors with tongued and grooved boards or sheets of plywood or chipboard at least 15 thick, joists at least 37 wide and a ceiling of 12.5 mm plasterboard with 5 neat gypsum plaster finish or tongued and grooved boards or sheets of plywood or chipboard on joists at least 37 wide with a ceiling of 12.5 mm plasterboard with joints taped and filled, will both have a resistance to fire of half an hour.

Resistance to the passage of heat

Timber upper floors that are exposed to outside air, such as floors over car ports, have to be insulated against heat transfer to meet the 1989 revision of the requirements of The Building Regulations 1985. The maximum U value of exposed floors has to be 0.45, the same as external walls. Insulation between joists in the form of low density glass fibre or rockwool rolls, mats or quilts can be laid on to the ceiling finish or semi-rigid batts, slabs or boards of fibre glass or rockwool which are friction fitted between joists and supported by nails or wood battens nailed to the sides of joists. To avoid cold bridges the insulation must extend right across the floor in both directions up to surrounding walls.

Resistance to airborne and impact sound

A boarded timber floor with a rigid plasterboard ceiling affords poor resistance to the transmission of airborne sound and impact sound.

The traditional method of insulating timber floors against sound was to spread a layer of plaster or sand on rough boarding fixed between the joists or sand on expanded metal lath and plaster as illustrated in Fig. 158. The layer of plaster or sand was termed pugging. Pugging is effective in reducing the transmission of airborne sound but has little effect in deadening impact sound. To deaden impact sound it is necessary to lay some resilient material between the floor surface material and the structural floor timbers. The combination of a resilient layer under the floor surface and pugging between joists will effect appreciable reduction of impact and airborne sound. Where the floor boards are nailed directly to

joists, through the resilient layer, as illustrated in Fig. 158, much of the impact sound deadening effect is lost.

Approved Document E, giving practical guidance to the requirements of Schedule 1 to The Building Regulations 1985 for resistance to airborne and impact sound, includes three specifications for the construction of timber floors.

The first specification is for a platform floor and absorbent blanket between joists as pugging as illustrated in Figs 159 and 160. The platform floor is formed as a sandwich of tongued and grooved timber or wood chipboard, at least 18 thick, bonded to a sub-base of at least 19 thick plasterboard or material of the same weight, laid on a resilient layer of mineral fibre, density between 60 and 100 kg/m³ and at least 25 thick, spread over a floor base of timber or wood chipboard at least 12 thick which is nailed to the joists. This platform or sandwich floor is very effective in deadening impact sound due to the close contact of the floor surface with the resilient layer. The pugging in this first specification is of mineral fibre, at least 100 thick and with a density of at least 12 kg/m³, laid on top of a ceiling at least 30 thick of two layers of plasterboard with joints staggered.

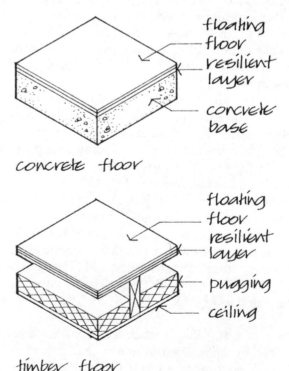

concrete floor

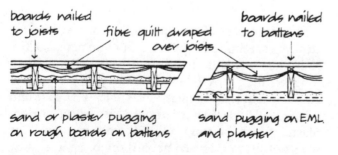

Sound insulation of timber floor

Fig. 158

timber floor

Fig. 159

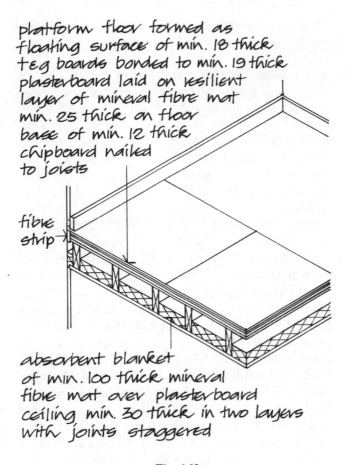

platform floor formed as
floating surface of min. 18 thick
teg boards bonded to min. 19 thick
plasterboard laid on resilient
layer of mineral fibre mat
min. 25 thick on floor
base of min. 12 thick
chipboard nailed
to joists

fibre
strip

absorbent blanket
of min. 100 thick mineral
fibre mat over plasterboard
ceiling min. 30 thick in two layers
with joints staggered

Fig. 160

The second and third specifications use what is called a ribbed floor of tongued and grooved timber or wood chipboard at least 18 thick, with the chipboard joints glued. In the second specification the boards are laid on or bonded to a base of plasterboard at least 19 thick or material of the same weight. The plasterboard base or boarded floor is nailed to battens, at least 50 nominal in width, that bear on to resilient strips between them and the top of each joist. The resilient strips are at least 25 thick with a density of 80 to 100 kg/m³. In these two specifications the resilient strips, laid continuously between the battens and the joists, deaden impact sound from the floating floor finish.

The pugging between joists in the second specification is of an absorbent blanket of mineral fibre, at least 100 thick and with a density of at least 12 kg/m³ laid on a ceiling of plasterboard at least 30 thick in two layers with joints staggered.

The pugging for the third specification is of dry sand or fine gravel with a weight of at least 80 kg/m²

laid in a ceiling of expanded metal lath covered with at least 19 thick dense plaster.

To provide a level floor surface it is necessary to fix the floor boards by nailing to battens or by fixing the boards to a firm level base of plasterboard or similar material, so that the boards can be cramped together. The boards are bonded to the plasterboard base with strips or pads of adhesive to keep them flat, as the tongues in the edges of the boards are cramped up into the grooves of adjacent boards, to produce a level floor finish. In this operation it is plainly simpler to cramp up the joints between large chipboards than the much less wide timber boards, which are only used when they are to be the finished floor surface. Joints between boards should be staggered as illustrated in Fig. 160.

To minimise flanking transmission of sound from the floor surface to the surrounding walls, a strip of resilient fibre material is fixed between the edges of platform and ribbed floors and surrounding solid walls as illustrated in Fig. 159, and a gap of at least 3 is left between skirtings and floating floors. To limit the transmission of airborne sound through gaps in the construction it is important to seal all gaps at junctions of wall and ceiling finishes and to avoid or seal breaks in the floor around service pipes.

The materials used as resilient layer in platform floors and as strips under ribbed floors, are fibre glass or rockwool in rolls or mat and the materials used as absorbent pugging between joists are fibre glass or rockwool in the form of rolls or semi-rigid slabs.

REINFORCED CONCRETE UPPER FLOORS

Reinforced concrete floors have a better resistance to damage by fire and can safely support greater superimposed loads than timber floors of similar depth. The resistance to fire, required by building regulations for most offices, large blocks of flats, factories and public buildings is greater than can be obtained with a timber upper floor so some form of reinforced concrete floor has to be used.

The following is a description of some of the concrete floors in use (see also Volume 4).

Strength and stability

Monolithic reinforced concrete floors: The word monolithic means one stone and is used in building to describe one unbroken mass of any material. A monolithic reinforced concrete floor is one unbroken solid mass of between 100 and 300 thick, cast in-situ concrete reinforced with mild steel reinforcing bars. To support the concrete while it is still wet and plastic, and for seven days after it has been placed, temporary centering has to be used. This takes the form of rough timber boarding, plywood, block-board, or steel sheets, supported on timber or steel beams and posts. The steel reinforcement is laid out on top of the centering and raised 20 above the centering by means of small blocks of fine concrete which are tied to the reinforcing bars with wire or by plastic spacers (see Volume 4). The wet concrete is then placed and spread on the centering, and it is compacted and levelled off. It is usual to design the floor so that it can safely span the least width of rooms and two opposite sides of the concrete are built into walls and brick partitions ½B each end or where the floor gives lateral support to walls it may be built in parallel to its span. Figure 161 illustrates a single monolithic concrete floor with part of the concrete taken away to show reinforcement and timber centering.

Centering: The temporary timber, plywood, blockboard or sheet steel support for monolithic concrete floor or roof is termed centering. The word centering was originally used for the timber form-work on which brick and stone arches and vaults were formed but today it is used to include the temporary support for concrete floors even though there is no curvature to the underside of the floor.

Reinforcement of concrete: A concrete floor has to carry loads just as a concrete lintel does and when loaded tends to bend in the same way. The steel reinforcing bars are cast into the underside of the floor with 20 of concrete cover below them to prevent the steel rusting and to give it one hour protection in case of fire. The thicker the concrete cover to reinforcement the greater the resistance of the floor to fire.

When the engineer designs a reinforced concrete floor he usually calculates the amount of steel reinforcement required for an imaginary strip of floor 300 wide spanning between walls, as though the floor were made up of 300 wide concrete beams placed side by side. The engineer will first calculate the combined superimposed and dead load that the floor has to support. The superimposed load is determined just as it is for timber floors and the dead load will include the actual weight of the concrete, the floor finish and the plaster on the soffit. From the loads and the span the required thickness of concrete will be determined and then the cross-sectional area of steel reinforcement for every 300 width of floor calculated. A rough method of determining the thickness of concrete required for floors of houses and flats is to allow 15 thickness of concrete for every 300 of span. The main reinforcement consists usually of 12 diameter mild steel rods spaced from 150 to 225

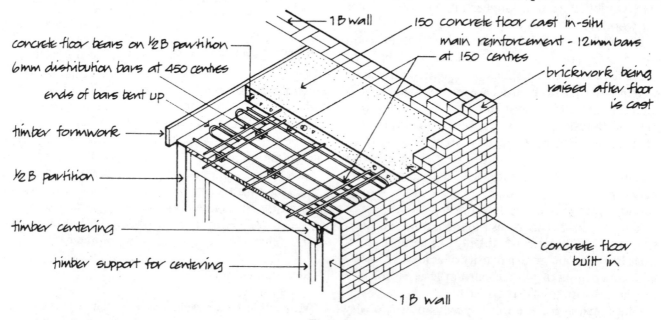

concrete floor bears on ½B partition

6mm distribution bars at 450 centres

ends of bars bent up

timber formwork

½B partition

timber centering

timber support for centering

1B wall

150 concrete floor cast in-situ
main reinforcement - 12mm bars
at 150 centres

brickwork being raised after floor is cast

concrete floor built in

1B wall

Fig. 161

apart, and these span across the floor between walls supporting the floor.

The 6 diameter mild steel rods wired across the main reinforcement are spaced at 450 to 900 apart and are called distribution rods or bars. These rods are tied to the main reinforcement with wire and keep the main reinforcing rods correctly spaced whilst the concrete is being placed and their main purpose is to assist in distributing point loads on the floor uniformly over the mass of the concrete.

In designing a reinforced concrete floor, as though it consisted of 300 wide beams, it is presumed that it bends in one direction only when loaded. In fact a monolithic concrete floor bends just as the skin of a drum does, when it is pressed in the middle. In presuming that the floor acts like a series of 300 wide beams the engineer can quite simply design it. But as no allowance is made for bending across the span, the floor as designed will be heavier and more expensive than it need be to safely carry its loads. The work involved in allowing for the bending that actually occurs in monolithic floors is considerably more than that required if it is presumed that the floor is a series of beams 300 wide.

Of recent years several firms have specialised in designing and constructing reinforced concrete and in order to be competitive their engineers make the more complicated calculation so as to economise in concrete and reinforcement.

Because the centering required to give temporary support to a monolithic concrete floor tends to obstruct and delay building operations 'self-centering' concrete floors are largely used today.

Self-centering concrete floors

These are constructed with precast reinforced concrete beams which are cast in the manufacturer's yard and are delivered to the building site where they are hoisted to the level of the floor and placed in position. Once in position they require no support other than the bearing of their ends on walls or beams. There are two types of self-centering reinforced concrete floors: (a) hollow beam floor units, and (b) solid 'T' section beams and hollow concrete infilling blocks.

(a) Hollow beam floor units: Hollow reinforced concrete beams, rectangular in section are precast. Figure 162 is a view of part of one beam. The width of the beams is usually 355, the depth 130 to 205 and

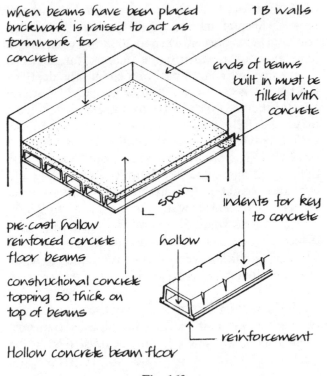

Hollow concrete beam floor

Fig. 162

the length up to 5.5. The concrete walls of these beams are from 15 to 20 thick and the steel reinforcement is cast into the angles of the beams.

The depth of the beams varies with the superimposed loads and the span. These beams are not suitable for floors carrying very heavy loads such as the floors of warehouses.

The beams are placed side by side with their ends bearing $\frac{1}{2}B$ on or into brick load bearing walls. If the ends of the beams are built into walls the ends should be solidly filled with concrete as the hollow beam is not strong enough to bear the weight of heavy brickwork.

The walls of the beams are made thin so that they are light in weight for transporting and hoisting into position. The thin walls of the beams are not strong enough to carry the direct weight of say furniture, and over them is spread a layer of concrete usually 50 thick which serves to spread point loads. The concrete is termed constructional concrete topping, and it is an integral part of this floor system. The concrete is mixed on the building site and is spread and levelled on top of the beams. In Fig. 162 a view of a hollow beam floor is shown. A hollow beam floor is lighter in weight than a similar monolithic concrete floor, but is deeper.

(b) Solid precast 'T' section beams with hollow light-weight concrete infilling blocks: Solid reinforced concrete beams generally shaped like an inverted 'T' in section are precast in the manufacturer's yard. Figure 163 is a view of part of a beam. The depth of the beams varies from 130 to 250 and they are 90 wide at the bottom. The beams are made in lengths up to 6.0. Hollow precast lightweight infilling blocks are made. These blocks are usually 225 wide and 225 or 300 long. They are made with one of the lightweight aggregates for lightness in handling and to reduce weight of the finished floor. The beams are placed at 270 centres with their ends bearing $\frac{1}{2}B$ on or into brick walls. The hollow blocks are then placed between the beams and the floor is finished with a 50 thick layer of constructional concrete topping. Figure 163 is a view of part of one of these floors.

The advantage of this floor over the hollow beam type is that its units can be handled by two men whereas hollow beams can only be hoisted by lifting gear.

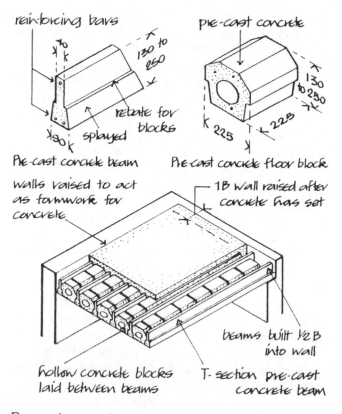

Fig. 163

Cold rolled steel deck and concrete floor

Of recent years profiled cold rolled steel decking, as permanent formwork and the whole or a part of the reinforcement to concrete has become the principal floor system for structural steel framed buildings as described in Volume 4. This floor system can be used with load bearing walls providing the hollow profile is filled at bearing ends.

Situ-cast reinforced concrete and terra cotta block floor

The resistance to damage by fire of a reinforced concrete floor depends on the protection, or cover, of concrete underneath the steel reinforcement. Under the action of heat, concrete is liable to expand and come away from its reinforcement. If, instead of concrete, pieces of burned clay tile are cast into the floor beneath the reinforcing bars the floor has a better resistance to fire than it would have with a similar thickness of concrete.

The particular advantage of this type of floor is its good resistance to damage by fire, and it is sometimes termed 'fire-resisting reinforced concrete floor'.

To keep the dead weight of the floor as low as possible, compatible with strength, it is constructed of situ-cast reinforced concrete beams with hollow terra cotta infilling blocks cast in between the beams. The words terra cotta mean 'earth burned'. The words terra cotta are used in the building industry to describe selected plastic clays which contain in their natural state some vitrifying material. After burning, the clay has a smooth hard surface which does not readily absorb water. The blocks are made hollow so that they will be light in weight and the smooth faces of the blocks are indented with grooves during moulding, to give a good 'key' for plaster and concrete. A typical T.C. (terra cotta) block is shown in Fig. 164. This type of floor has to be given temporary support with timber or steel centering. The T.C. blocks and the reinforcement are set out on the centering, and pieces (slips) of clay tile are placed underneath the reinforcing bars. Concrete is then placed and compacted between the T.C. blocks and spread 50 thick over the top of the blocks. Figure 164 is a view of part of one of these floors.

The floor is built into walls $\frac{1}{2}B$ as shown. This type of floor can span up to 5.0 and the depth of the blocks, the depth of the finished floor and the size

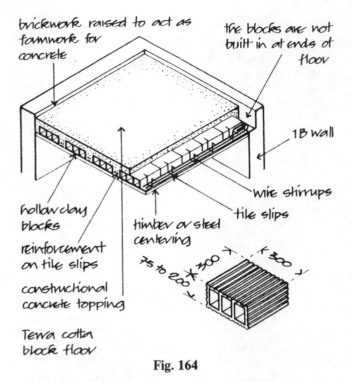

brickwork raised to act as formwork for concrete

the blocks are not built in at ends of floor

1B wall

wire stirrups

tile slips

hollow clay blocks

reinforcement on tile slips

timber or steel centering

constructional concrete topping

75 to 200   300   300

Terra cotta block floor

Fig. 164

and number of reinforcing bars depend on the superimposed loads and span. This type of floor is much less used today than it was because of the considerable labour in placing the hollow T.C. pots and reinforcement and because of the need for temporary centering.

Floor finishes to concrete upper floors

The floor finishes described for solid concrete ground floors may be used on similar concrete or screed bases, for upper floors.

Fire resistance

The resistance to fire of a reinforced concrete floor depends on the thickness of concrete cover to steel reinforcement, as the expansion of the steel under heat will tend to cause the floor to crack and ultimately give way. The practical guidance given in Approved Document B to The Building Regulations 1985 specifies at least 20 cover of concrete for reinforcement to give a notional period of fire resistance of one hour.

Resistance to the passage of heat

A suspended, upper concrete floor that is exposed to outside air has to be insulated against excessive

transfer of heat, to meet the 1989 revision of the requirements of The Building Regulations 1985, so that the construction has a maximum U value of 0.45. The most practical way of insulating this type of floor construction is to form a layer of insulating material under a screeded or platform boarded floor similar to that described for suspended concrete ground floors.

Resistance to airborne and impact sound

The weight of a solid concrete floor is generally sufficient to reduce intrusive airborne sound but provides poor resistance to impact sound. In Approved Document E, giving practical guidance to meeting the requirements of The Building Regulations 1985 for the reduction of airborne and impact sound through concrete floors, are details of solid floors with resilient covering materials such as carpet and solid floors with floating floor finish.

A concrete floor with its screed and plaster finish, weighing at least 365 kg/m^2, will act as an effective barrier to airborne sound. To deaden impact sound this type of solid floor should be covered with carpet having a thickness of 4.5 mm, including resilient underlay.

Concrete floors, with screed and plaster finish, weighing at least 220 kg/m^2 should have a floating floor finish to act with the concrete base in reducing airborne sound and to deaden impact sound. The two floating floors are wood-based and screed-based.

A floating wood-based floor is formed as a platform of tongued and grooved timber boards, wood chipboard or plywood at least 18 thick nailed to timber battens, which is laid on top of mineral wool mat at least 13 thick with a density of at least 36 kg/m^3 as illustrated in Fig. 159. The rolls of acoustic insulation are laid across the top of the concrete floor with the edges of rolls tightly butted. The timber platform or raft is fabricated in panels of convenient size for handling and the edges of panels are butted together by wedging to surrounding walls and laid with end joints staggered. To minimise flanking transmission of sound from the floor to surrounding walls, there should be a resilient strip between the edges of the platform floor and surrounding walls.

A floating screed-based floor is formed with an acoustic layer of compressed mineral fibre or expanded polystyrene boards laid on the concrete floor on which a screed is spread and trowelled to a

CONSTRUCTION OF BUILDINGS

level finish ready for the floor finish. The top of the insulation layer should be covered with building paper to prevent screed running into the insulation. The screed should be at least 65 thick and reinforced with wire mesh to control cracking. To minimise flanking transmission the insulation should be continuous between the edges of the screed and the surrounding walls.

CHAPTER FOUR

ROOFS

FUNCTIONAL REQUIREMENTS

The functional requirements of a roof are:

Strength and stability
Resistance to weather
Durability and freedom from maintenance
Fire resistance
Resistance to the passage of heat
Resistance to airborne sound

Strength

The strength of a roof depends on the characteristics of the materials from which it is constructed and the way in which they are put together in the form of a flat platform or some form of triangulated frame.

Pitched roofs depend for strength on the triangulated form of the roof structure, in which the considerable depth of the roof at mid span gives it strength and for which comparatively slender sections can safely be used. The tables in Approved Document A for pitched roofs give sizes of members for roofs pitched at from 10 to $42\frac{1}{2}$ degrees, for small buildings.

A flat roof acts in much the same way as a floor, as a level platform designed to support loads. The loads that a roof supports, wind pressure or uplift and snow, are generally less than those supported by a floor. In Approved Document A, giving practical guidance to the requirements of Schedule 1 to The Building Regulations 1985 for small domestic buildings, are tables of the sizes of joists required for flat roofs, related to loads and span. There are four tables. The first two give joist sizes for roofs with access only for maintenance and repair. The last two tables give joist sizes for roofs with access not limited to maintenance and repair and, therefore, liable to heavier loads, such as use as a terrace.

Stability

A roof is constructed to support the dead load of the roof structure and its covering, insulation and internal finishes, snow loads and pressure or suction due to wind without undue deflection or distortion. The dead load can be calculated from the unit weight of materials set out in BS 648. Snow loads are assumed from average snow falls. The pressure of wind on a roof will depend on the exposure, height and shape of the roof and surrounding buildings. Wind blowing across a roof will tend to cause pressure on the windward and suction on the opposite side of the building. The actual pressure of wind on buildings is very difficult to predict with any certainty.

A roof may be constructed as a flat roof, that is a timber, metal or concrete framed platform which is either horizontal or inclined up to five degrees to the horizontal, as a sloped or sloping roof inclined over five degrees and up to ten degrees or as a pitched roof with one or more slopes pitched at more than ten degrees to the horizontal as illustrated in Fig. 165.

A simple pitched roof has equal slopes rising, that is pitched, to a central ridge with horizontal ties at ceiling level as illustrated in Fig. 165. The stability of a pitched roof depends on the depth of the triangular framing at mid span and it is this depth that gives a pitched roof its stability across the span of the roof. There is an inherent instability across the slopes, parallel to the ridge, which may allow the triangulated frames to rack or fall over like a stack of books on a shelf, that will fall when not supported at ends. Roofs are braced against racking by end gable walls, by hipped ends or by cross bracing by diagonal roof boarding or cross braces.

A flat roof is constructed in the same way as a floor as either a timber or concrete platform. The stability of a flat roof depends on adequate support from walls or beams and sufficient depth or thickness of

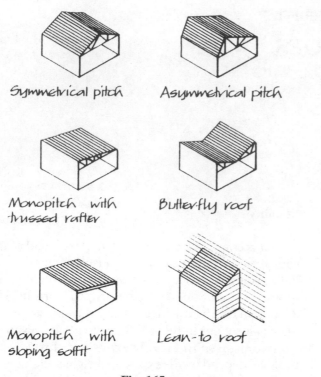

Symmetrical pitch   Asymmetrical pitch

Monopitch with trussed rafter   Butterfly roof

Monopitch with sloping soffit   Lean-to roof

Fig. 165

timber joists or concrete relative to spans, and assumed loads to avoid gross deflection under load. The flat roof type, termed a monopitch roof, may be constructed as a sloping platform with a sloping soffit or as a triangular frame as illustrated in Fig. 165, with the ceiling flat and the roof sloping. This roof acts structurally as a pitched roof, as its stability depends on the depth of the triangular frame at mid span.

The butterfly roof illustrated in Fig. 165 is in effect two monopitch roofs which depend for support and stability on a central supporting beam or wall.

Resistance to weather

A roof excludes rain through the material with which it is covered, varying from the continuous impermeable layer of asphalt covering that can be laid horizontal to exclude rain, to the small units of clay tiles that are laid overlapping down slopes so that rain runs rapidly to the eaves. In general, the smaller the unit of roof covering, such as tile or slate, the greater must be the pitch or slope of the roof to exclude rain that runs down in the joints between the tiles or slates on to the back of another tile or slate lapped under and so on down the roof. Larger units such as profiled sheets (see Volume 3) can be laid at

pitches less than that required for tiles. Impermeable materials such as asphalt and bitumen that are laid without joints can be laid flat and sheet metals such as lead and copper that are joined with welts can be laid with a very shallow fall.

A roof structure will be subject to movements due to variations in loading by wind pressure or suction and snow loads and movements due to temperature and moisture changes. The great advantage of the traditional roofing materials slate and tile, is that as the small units are hung, overlapping down the slope of roofs, the very many open joints between the tiles or slates can accommodate movements in the roof structure without breaking slates or tiles or letting in rainwater, whereas large unit size materials and continuous roof coverings may fail if there is inadequate provision of movement joints.

Durability and freedom from maintenance

The durability of a roof depends largely on the ability of the roof covering to exclude rain and snow. Persistent penetration of water into the roof structure may cause or encourage decay of timber, corrosion of steel or disintegration of concrete.

The traditional materials slates and tiles when laid at an adequate pitch (slope) and properly double lapped to exclude rain will, if undisturbed, have a useful life of very many years and will require little if any maintenance and may survive the anticipated life of buildings. Because of the variations in shape, colour and texture of natural slates and handmade clay tiles it is generally accepted that these traditional materials are often an initial and continuing attractive feature of buildings. The uniformity of shape, colour and texture of machine made slates and tiles have, by common acceptance, a less pleasing appearance.

The non-ferrous sheet metal coverings, lead, copper, zinc and aluminium overlapped and jointed and fixed to accommodate movement and with a slope adequate for rainwater to run off, will have a useful life of many years and should require little if any maintenance during the life of most buildings. As these durable roof coverings are not a familiar part of the appearance of the majority of buildings in this country, there is no broad consensus of opinion as to their effect on the appearance of buildings.

The flat roof materials asphalt and bitumen felt are by their nature short to medium term life materials due to the gradual oxidisation and hardening of the material which has a useful life of some 20 to 30 years at most. While the covering to a flat roof is not generally one of the more visible features of buildings, it is generally accepted that these roofing materials do not have an attractive appearance.

Fire resistance

The requirements for the control of spread of fire in Schedule 1 to The Building Regulations 1985 for dwelling houses, limit roof construction relative to the proximity of boundaries of the site of a building, by reference to the materials of roof covering. The traditional materials, slates, tiles, non-ferrous sheet metals and also mastic asphalt do not encourage spread of flame and there is, therefore, no limitation to their use. Bitumen felt roof coverings by themselves, do to an extent encourage spread of flame unless they are covered with stone chippings at least 12.5 mm deep, non-combustible tiles, sand and cement screed or macadam.

Resistance to the passage of heat

The materials of roof structures and roof coverings are generally poor insulators against the transfer of heat and it is usually necessary to use some material which is a good insulator, such as lightweight boards, mat or loose fill to provide insulation against excessive loss or gain of heat. The 1989 revision of the requirements of The Building Regulations 1985 for the insulation of roofs of dwellings is a maximum U value of 0.25.

The most economical method of insulating a pitched roof is to lay or fix some insulating material between or across the ceiling joists, the area of which is less than that of the roof slope or slopes. This insulating layer will at once act to reduce loss of heat from the building to the roof space, and reduce gain of heat from the roof space to the building. As the roof space is not insulated against loss or gain of heat it is necessary to insulate water storage cisterns and pipes in the roof against possible damage by freezing. Where the space inside a pitched roof is used for storage or as part of the building it is usual to insulate the roof slopes.

With insulation at ceiling level in a pitched roof it is necessary to ventilate the roof space against moisture vapour that might encourage decay of timbers. In warmer climates than the United Kingdom it is often practice to provide ventilation to the roof to reduce the temperature inside the roof.

With the insulation at ceiling level the roof will be roughly at outside air temperature. This is a form of cold construction, or a cold roof. With the insulation fixed across the roof rafters, under the roof covering, the roof is insulated against changes in outside air temperature and is a form of warm construction or a warm roof. Obviously the roof space below insulation across roof rafters should not be ventilated to outside air.

Insulating materials may be applied to the underside or top of flat roofs or between the joists of timber flat roofs. Rigid structural materials, such as wood wool slabs, that serve as a roof deck and rigid insulation boards are laid on the top of the roof and non-rigid materials either between joists or on top of the roof below some form of decking.

The position for insulation will be affected by the type of flat roof structure, the nature of the insulation material, the most convenient place to fix it and the material of the roof finish. The most practical place to fix the insulation is on top of the timber or concrete roof deck or structure, under the roof covering.

With insulation on top of the roof deck the structure will be insulated from outside air and maintained at roughly the inside air temperature. This arrangement is sometimes described as warm construction, as the roof structure is as warm as the inside of the building, or the roof is said to be a warm roof. The advantage of the warm roof, particularly with timber roofs, is that there is no necessity to ventilate the roof itself against condensation moisture. The very considerable disadvantage of the warm roof is that as the insulation is directly under the roof covering, the material of the covering will suffer very considerable temperature fluctuations between hot sunny days and cold nights, as the roof structure will absorb little if any of the heat or cold that the roof covering is subjected to. These very considerable temperature fluctuations will cause tar and bitumen coverings, such as asphalt and felt, to oxidise more rapidly, become brittle and fail and cause severe mechanical strain in other coverings. An inverted or upside down warm roof, with the insulation on top of the roof covering will protect the roof covering from severe temperature fluctuations.

With the insulation below, the roof structure is subject to the fluctuations of temperature between hot sunny days and cold nights and the construction is sometimes referred to as cold construction or cold roof. The disadvantage of cold construction or cold roof is that the moisture vapour pressure of warm inside air may cause vapour to penetrate the insulation and condense to water on the cold side of the insulation where it may cause corrosion of reinforcement in concrete and promote decay in a timber roof. Cold roofs should be protected by a vapour check on the warm side of the insulation and timber roofs ventilated against build up of moisture.

Vapour check, vapour barrier

To reduce the movement of warm moist air from inside a building to the cold side of insulation, it is practice to fix a layer of some impermeable material, such as polythene sheeting, to the underside, the warm side, of insulation that is permeable to moisture vapour, as a check to the movement of moisture vapour. By definition a vapour check is any material that is sufficiently impermeable to check the movement of moisture vapour without being an impenetrable barrier. The term vapour barrier is sometimes unwisely used in lieu of the term vapour check. The use of the word barrier suggests that the material serves as a complete check to the movement of vapour. The practical difficulties of forming an unbroken barrier between sheets of polythene at the junction of walls and ceiling and around rooflights and pipes suggest the term vapour check is more appropriate.

Some insulating materials, such as the organic, closed cell boards, for example extruded polystyrene, are substantially impermeable to moisture vapour. When these boards can be close butted together or provided with rebated or tongued and grooved edges and cut and close fitted to junctions with walls and around pipes, they will serve as a vapour check and there is no need for an additional layer of vapour check material.

Resistance to airborne sound

The resistance of a roof to the penetration of airborne sound is not generally considered unless the building is close to a busy airport. The mass of the materials of a roof is the main consideration in the reduction of airborne sound. A solid concrete roof will more effectively reduce airborne sound than a similar timber roof. The introduction of mineral fibre slabs, batts or boards to a timber roof will have some effect in reducing intrusive, airborne sound.

PITCHED ROOFS

Strength and stability

A pitched roof has roof slopes at a pitch or slope of more than 10 degrees to the horizontal. The most common roof shape is the symmetrical pitch roof pitched to a central ridge with equal slopes as illustrated in Fig. 165. A monopitch roof has one slope as illustrated in Fig. 165, which is a free standing version of the lean-to roof. Variants of these two simple shapes are the asymmetrical pitch and the butterfly roof as illustrated in Fig. 165. Figure 166 illustrates the terms used to describe the parts of a pitched roof.

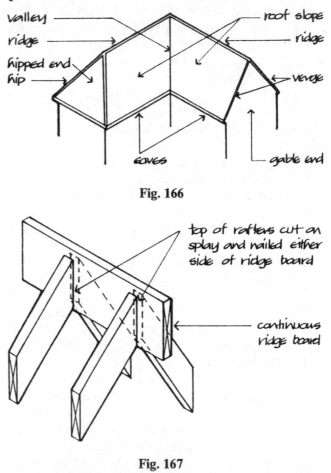

Fig. 166

Fig. 167

The traditional English roofing materials, slate and tile can only successfully be used if fixed on to a surface inclined at least 25 degrees to horizontal. The traditional way of constructing a roof with sloping surfaces is to pitch timber rafters either side of a central horizontal ridge, as illustrated in Fig. 167 with the rafters bearing on a wall plate on the supporting walls as illustrated in Fig. 168. This, the simplest form of pitched roof, Fig. 169, is termed a couple roof because each pair of rafters acts like two arms pinned at the top and the mechanical term for such an arrangement is a couple. The weight of the roof tends to spread the rafters of a couple roof and overturn the supporting walls, as illustrated in Fig. 170, and the span, or horizontal distance between wall plates, is limited to 3.5.

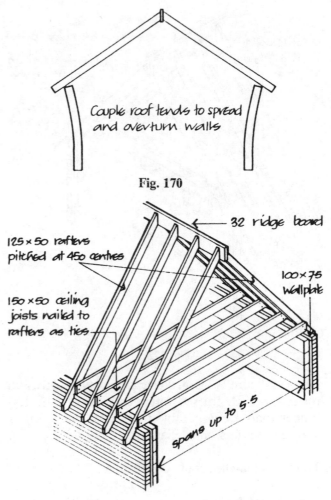

Couple roof tends to spread and overturn walls

Fig. 170

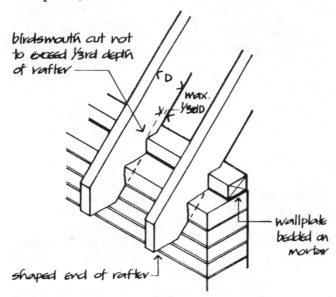

birdsmouth cut not to exceed 1/3rd depth of rafter

D

max. 1/3rd D

wallplate bedded on mortar

shaped end of rafter

Fig. 168

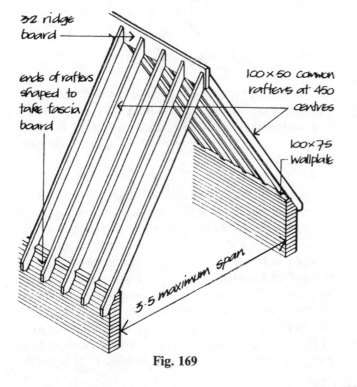

32 ridge board

ends of rafters shaped to take fascia board

100 × 50 common rafters at 450 centres

100 × 75 wallplate

3.5 maximum span

Fig. 169

125 × 50 rafters pitched at 450 centres

150 × 50 ceiling joists nailed to rafters as ties

32 ridge board

100 × 75 wallplate

spans up to 5.5

Close couple roof

Fig. 171

In the traditional pitched roof form, timber ties are nailed to the foot of pairs of rafters to prevent them spreading under the load of the roof. These ties may also serve to support the ceiling finish. This type of roof, illustrated in Fig. 171, is termed a close (or closed) couple roof.

A modification of the close couple roof is the collar roof, where the ties are fixed between pairs of rafters, one-third the height of the roof up from the wall plate, as illustrated in Fig. 172, so that rooms may extend up into part of the roof.

With stress grading of timber allowing more accurate sizing of structural timber, the use of connector plates and factory prefabrications, many timber-framed pitched roofs today are constructed as trussed rafters. A trussed rafter is a triangular roof frame of rafters, ceiling joist and internal webs joined with spiked connector plates and assembled in a

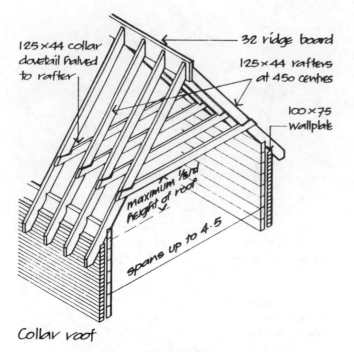

Collar roof

Fig. 172

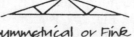
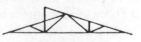

Trussed rafter types

Fig. 173

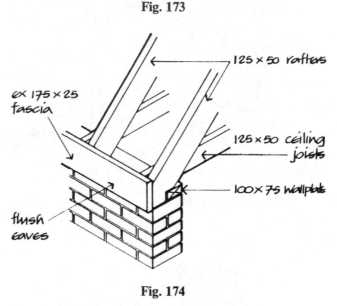

Fig. 174

factory. Figure 173 is an illustration of typical trussed rafter types. A trussed rafter uses up to 60% less timber than a comparable traditional pitched roof and requires less on-site labour.

Traditional pitched roof construction

Size of roof timbers: Rafters are usually 38 to 50 thick, 100 to 150 deep and are spaced at from 400 to 600 apart measured from the centre of one rafter to the centre of the next. The depth of rafters and the centres at which they are fixed depends on the type and weight of the roof coverings they have to support and their unsupported length. In addition to the dead weight of the roof covering, such as tiles or slates, the rafters have to be able to resist the pressure of wind. The superimposed loads which roofs must be able to support are given in kilonewtons per square metre (kN/m^2) of surface area of roof and are calculated to allow for the maximum pressure a roof is likely to have to withstand due to wind and the weight of snow on the roof.

Ceiling joists (ties) are 38 to 50 thick and from 75 to 225 deep and since they are nailed to the sides of the rafters they will be spaced at the same centres as the rafters. Collars are usually 44 thick and are usually as deep as roof rafters. The ridge board is usually 25 to 38 thick and so deep that the whole depth of the splay cut ends of rafters bear on it. The

depth of the ridge board will depend on the depth of the rafters and the angle at which they are pitched. The wall plate is usually 100 wide and either 75 or 50 deep and is bedded true level in mortar on the brickwork.

Eaves is a general term used to describe the lowest courses of the slates or tiles and the timber supporting them. The eaves of most pitched roofs are made to project some 150 to 300 beyond the external face of walls. In this way the roof gives some protection to the walls, and enhances the appearance of a building. The eaves of the roof of sheds, outhouses and other outbuildings are sometimes finished flush with the face of the external wall of the building. The purpose of this is to economise in timber roof covering. The construction of flush eaves is illustrated in Fig. 174. It

140

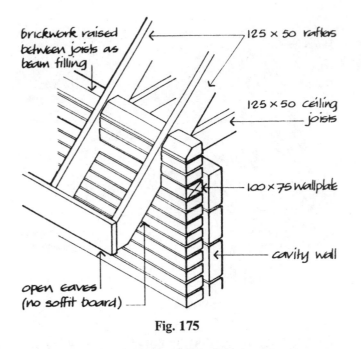

brickwork raised between joists as beam filling

125 × 50 rafters

125 × 50 ceiling joists

100 × 75 wallplate

cavity wall

open eaves (no soffit board)

Fig. 175

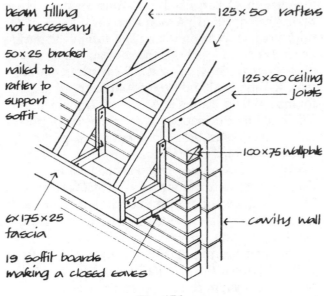

beam filling not necessary

50 × 25 bracket nailed to rafter to support soffit

6 × 175 × 25 fascia

19 soffit boards making a closed eaves

125 × 50 rafters

125 × 50 ceiling joists

100 × 75 wallplate

cavity wall

Fig. 176

will be seen that the ends of the rafters are cut so that they are flush with the wall face and a timber gutter (or fascia) board is then nailed to them. The roof covering drains to an eaves gutter fixed to the fascia board. A building with flush eaves and pitched roof looks unfinished as though it were wearing a cap too tight for it.

Projecting eaves are constructed by making the ends of the roof rafters project as illustrated in Fig. 175. It will be seen that the ends of the rafters are cut so that a gutter board can be nailed to them.

Figure 175 is an illustration of an open eaves roof, so called because the underside of the rafters is exposed. To prevent wind blowing up into the roof between the rafters, beam filling is formed between them. This beam filling consists of brickwork raised between the rafters up to the underside of the roof covering.

Usually the soffit of projecting eaves is closed by fixing boards or sheets on it and this construction is illustrated in Fig. 176. It will be seen that short lengths of small section sawn timber are nailed to the side of each rafter to form a bracket to which the boarding of the soffit can be securely fixed. Timber boarding 16 or 19 thick or plywood is usually employed for the soffit. It will be seen from Fig. 176 that the wall plate is bedded on the external leaf of the wall so that the level of the soffit of the eaves is higher than it would be were the plate bedded on the inner leaf, so that the head of windows may be close to ceiling level.

Purlin or double roof: Another way of constructing pitched roofs with spans of from 4.5 to 7.5 is to form a purlin roof. A purlin is a continuous timber fixed horizontally under the roof rafters to give them support between the ridge and the wall plate. The purlin in turn is supported by means of timber struts which bear on to a load bearing partition. The arrangement of these timbers is shown in Fig. 177. It will be seen that purlins support the rafters midway between the ridge and eaves and that they are supported by struts at intervals of about 1.8 along

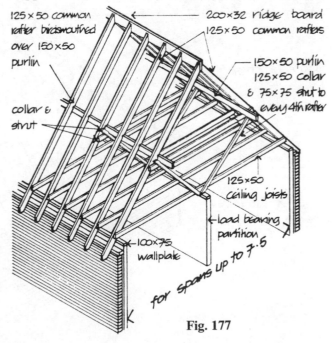

125 × 50 common rafter birdsmouthed over 150 × 50 purlin

collar & strut

200 × 32 ridge board

125 × 50 common rafters

150 × 50 purlin

125 × 50 collar & 75 × 75 strut to every 4th rafter

125 × 50 ceiling joists

load bearing partition

100 × 75 wallplate

for spans up to 7.5

Fig. 177

141

their length. Collars fixed every fourth rafter serve to brace the roof and provide a secure fixing for the purlins which bear on them. The size of the purlins depends on the weight of the roof and their unsupported length between struts. With struts not more than 1.8 apart a 175 deep by 50 thick purlin is used for most rafters. Tables in Approved Document A to The Building Regulations 1985 relate size of purlins to load and span between supports to purlins for houses and small buildings.

Collars of the same section as the roof rafters are fixed to every third or fourth rafter. Struts are usually 75 square in section. The foot of the struts is fixed to a timber wall plate bedded in mortar on the load bearing partition.

The load bearing partition does not have to be central between the external walls in order to give satisfactory support to the struts which in turn support purlins. If there are load bearing partitions running at right angles to the ridge they can be used to support struts.

The pitched roofs of terraced houses are commonly constructed with purlins the ends of which can be supported by walls dividing the houses. To prevent the spread of fire from one house to another,

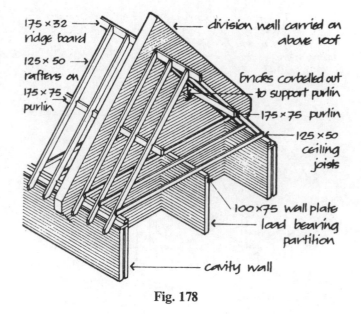

Fig. 178

the Building Regulations require that the wall dividing houses, the party or separating wall, be carried up to at least the level of the underside of the roof covering and purlins can be supported on brick or stone or concrete corbels built out from these walls. The construction is illustrated in Fig. 178.

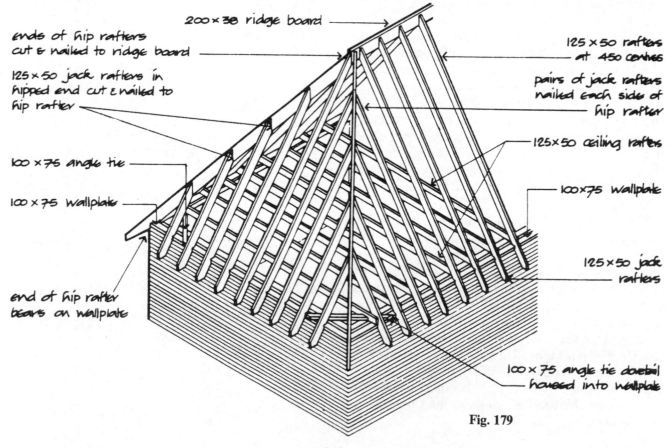

Fig. 179

142

Hips: The most economical way of constructing a pitched roof is to form it with two slopes with gable ends. But a simple gable end roof sometimes looks clumsy due to the great area of tile or slate covering and this can be avoided by forming hipped ends to the roof.

The hipped ends of roofs are pitched at the same slope as the main part of the roof and the rafters in the triangle of the hipped end are pitched up to a hip rafter. The hip rafters carry the ends of the cut rafters in the hipped end and those of the main roof slopes. The hip rafter is usually 38 thick and 200 to 250 deep. The cut 'jack rafters' are nailed each side of the hip rafter as shown in Fig. 179.

Because the hip rafter carries the ends of several jack rafters it tends to overturn the walls at the corner of the building where it bears on the wall plates and to resist this an angle tie should always be fixed across the angle of the roof as shown in Fig. 179.

The angle ties are usually 100×75 timber and are either firmly bolted to or dovetail housed into the top of the wall plates some 600 from the corner of the building. The dovetail housed joint is similar to that between a collar and rafters used in a collar roof.

One of the advantages of a hipped end to roofs is that the hipped end, being pitched at right angles to the main roof, gives stability to the roof.

Trussed roof construction: The span of a close couple roof is limited to 5.5, which is less than the width of most buildings. A purlin roof which can span up to 7.5, depends for support on load bearing partitions conveniently placed and these partitions often restrict freedom in planning the rooms of buildings.

A method of constructing pitched roofs, with spans adequate to the width of most small buildings, that that do not need intermediate support from internal partitions is to form timber trusses. The word truss means tied together and a timber roof truss is a triangular frame of timbers securely tied together. The traditional timber roof truss was constructed with large section timbers that were cut and joined with conventional mortice and tenon joints that were strapped with iron straps screwed or bolted to the truss. The traditional King post and Queen post trusses supported timber purlins and rafters.

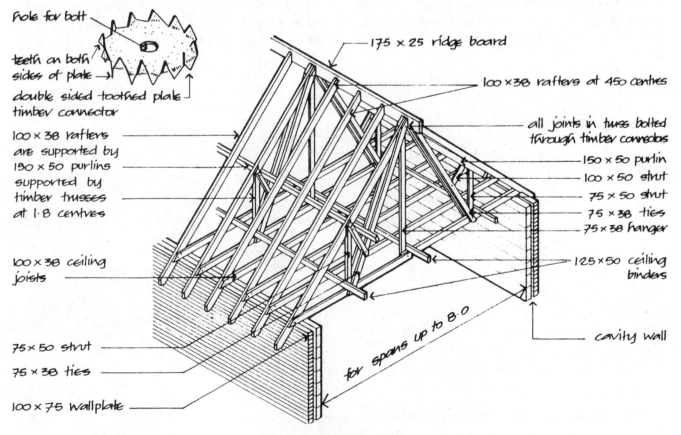

hole for bolt

teeth on both sides of plate →

double sided toothed plate timber connector

100×38 rafters are supported by 150×50 purlins supported by timber trusses at 1·8 centres

100×38 ceiling joists

75×50 strut

75×38 ties

100×75 wallplate

175×25 ridge board

100×38 rafters at 450 centres

all joints in truss bolted through timber connectors

150×50 purlin

100×50 strut

75×50 strut

75×38 ties

75×38 hanger

125×50 ceiling binders

cavity wall

for spans up to 8.0

Fig. 180

143

The combination of shortage of timber that followed the end of the Second World War (1945) and the need for greater freedom in planning internal partitions prompted the development of the economical timber truss designed by the Timber Development Association. The timbers of the truss were bolted together with galvanised iron timber connectors, bolted between timbers at connections. The strength of the truss was mainly in the rigidity of the connection. The trusses, spaced at 1.8 centres, supported purlins and roof rafters as illustrated in Fig. 180. To economise in the size of ceiling rafters, timber runners and hangers gave intermediate support. These trusses were used for many years as the most economical and satisfactory method of framing pitched roofs for small buildings.

For the sake of economy in site labour and timber the trussed rafter was first used in this country in 1964. A trussed rafter is fabricated from light section, stress graded timbers that are accurately cut to shape, assembled and joined with galvanised steel connector plates. Much of the preparation and fabrication of these trussed rafters is mechanised, resulting in accurately cut and finished rafters that are delivered to site ready to be lifted and fixed as a roof frame with the minimum of site labour. The members of the truss are joined with steel connector plates with protruding teeth that are pressed into timbers at connections to make a rigid joint. Trussed rafters, that serve as rafters and ceiling joists, are fixed at from 400 to 600 centres as illustrated in Fig. 181. The trussed rafters are fixed to wall plates. As the rafters are trussed there is no need for a ridge board and the eaves may be flush or projecting.

Since they were first introduced into this country, these trussed rafters have been every extensively used for houses, particularly on housing estates of similar buildings where repetitive production provided the greatest economy.

A consequence of the pressure for economy by the use of slender timber sections for trussed rafters spaced at maximum centres and careless workmanship, has been the failure of some trussed rafter roofs due to the inherent instability of a pitched roof parallel to the ridge and buckling of the slender section trussed rafters.

Approved Document A, giving practical guidance to meeting the requirements of Schedule 1 to The Building Regulations 1985 for small buildings, states that for stability, trussed rafter roofs should be braced to the recommendations in BS 5268: Part 3.

Because the members of a trussed rafter roof are delivered to site as framed, triangulated units, it has been common to employ unskilled labour in the

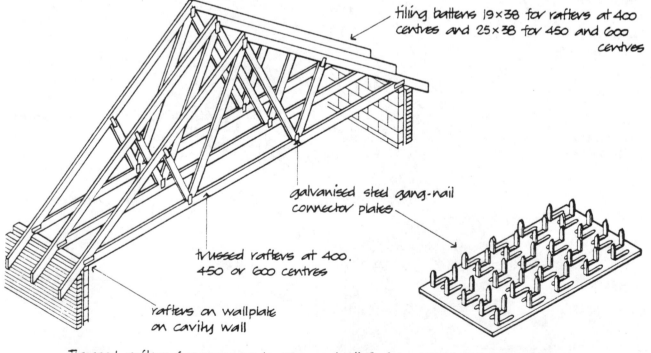

tiling battens 19×38 for rafters at 400 centres and 25×38 for 450 and 600 centres

galvanised steel gang-nail connector plates

trussed rafters at 400, 450 or 600 centres

rafters on wallplate on cavity wall

Trussed rafters for spans up to 12m and pitch from 15° to 40°

Fig. 181

erection of these roofs. The result has been that trusses, which were often not accurately spaced nor erected and fixed true vertical, have buckled under load. To take account of practical factors regarding the use of trussed rafters on site and the limit of experience of their behaviour under test and in service, there are recommendations in BS 5268: Part 3 for their use, erection and stability bracing. For the roofs of domestic buildings the members of trussed rafters with spans up to 11.0 should be not less than 35 thick and those with spans of up to 15.0, 47 thick.

The slender section trusses are liable to damage in storage and handling. They should be stored on site either horizontal on a firm level base or in a vertical position with adequate props to avoid distortion. In handling into position each truss should be supported at eaves rather than mid span to avoid distortion. The trussed rafters should be fixed vertical on level wall plates at regular centres and maintained in position with temporary longitudinal battens and raking braces. The rafters should be fixed to wall plates with galvanised steel truss clips that are nailed to the sides of trusses and wall plates. A system of stability bracing should then be permanently nailed to the trussed rafters. The bracing that is designed to maintain the rafters in position and to reduce buckling under load, is illustrated in Fig. 182. The bracing members should be 25×100 and nailed with two 3.35mm $\times$ 75 galvanised, round wire nails at each cross-over.

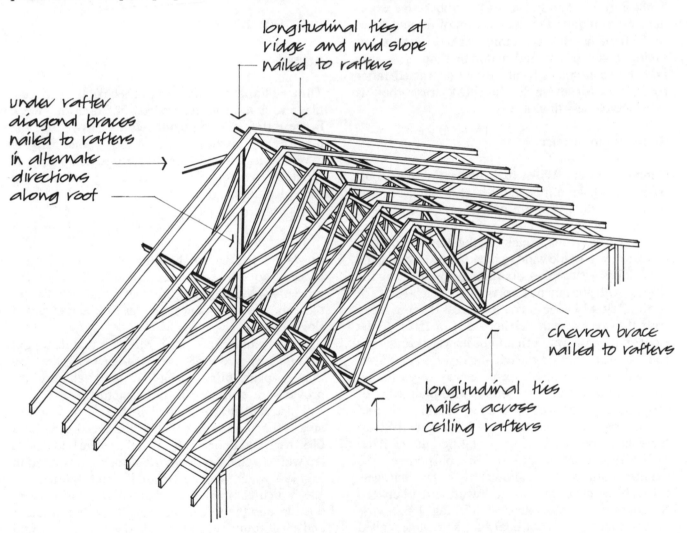

longitudinal ties at ridge and mid slope nailed to rafters

under rafter diagonal braces nailed to rafters in alternate directions along roof

chevron brace nailed to rafters

longitudinal ties nailed across ceiling rafters

Stability bracing to trussed rafter roof

Fig. 182

145

The longitudinal brace at the apex of trusses acts in much the same way as a ridge board and those halfway down slopes like purlins to maintain the vertical stability of trusses. The longitudinal braces, also termed binders, at ceiling level serve to resist buckling of individual trussed rafters. The diagonal, under rafter braces and the diagonal web braces serve to stiffen the whole roof system of trussed rafters by acting as deep timber girders in the roof slopes and in the webs.

The bracing shown in Fig. 182 is for stability. Depending on the position of exposure of the building, additional bracing may be necessary against wind pressures.

The implication of the recommendations in BS 5268: Part 3 is that more control and better workmanship is required in the erection of trussed rafter roofs than has been common. There is thus less saving in site labour and materials than previously and the economic advantage of the trussed rafter roof over, for example the TDA truss roof, is considerably less than it was.

Resistance to weather

Wind exclusion: When fuel for heating was comparatively cheap and standards of thermal comfort were lower than they are now, it was not usual to insulate roofs against transfer of heat. The roof space under pitched roofs, which to an extent acted as an insulator, was made wind tight either by covering the roof slopes with timber boarding or by covering the slopes with bitumen impregnated felt. Rolls of roofing felt, 813 wide are nailed across the roof rafter from the eaves with the felt lapped 75 or 150 up to the ridge so that any water that finds its way through the roof covering runs on the felt and down to the eaves. A more effective and expensive way of excluding wind is to cover the roof slopes with plain edge or tongued and grooved boarding, roofing felt and counter battens 38×19 or 50×25 that are fixed up the roof slopes with tiling or slating battens fixed across the counter battens. The purpose of the counter battens is to allow any water that gets through the covering to run down the felt between the counter battens as illustrated in Fig. 183.

An advantage of roof boarding, sometimes called sarking, is that it acts to brace the roof against the inherent instability of a pitched roof.

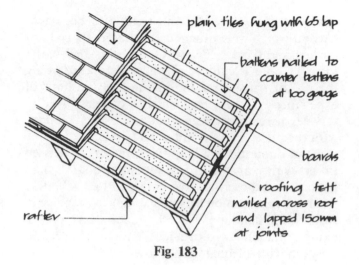

Fig. 183

PITCHED ROOF COVERINGS

Resistance to weather

The traditional covering for pitched roofs, plain clay tiles and natural slates, are less used than they were because they are comparatively expensive and the majority of pitched roofs of new buildings are covered with concrete tiles or fibre cement slates.

The small unit pitched roof coverings are single lap tiles, plain tiles and slates.

Tiles

Single lap tiles are so shaped that they overlap the edges of adjacent tiles in each course. This overlap prevents water entering the roof between adjacent tiles, and in consequence the tiles can be laid with a single end lap.

The advantage of single lap tiling is that its weight per unit of area of roof covered is up to 40% less than that of plain tiling of similar material and this provides for economy of roof structure.

Single lap tiles originated in the Mediterranean area as shaped clay tiles of half round section **Spanish tiles** overs and unders, Fig. 184, that were laid to side lap with a single lap down the slope as illustrated in Fig. 190, which were developed into the pantile, Fig. 184, which combined the channel and cover tile in one tile, and the **Italian tile**, Fig. 185, of a flat under and a half round section over, laid with a side lap and a single lap down the slope, as illustrated in Fig. 191, developed into the single Roman and double Roman tile, combining the under and over of the Italian tile in one tile, as illustrated in Fig. 185.

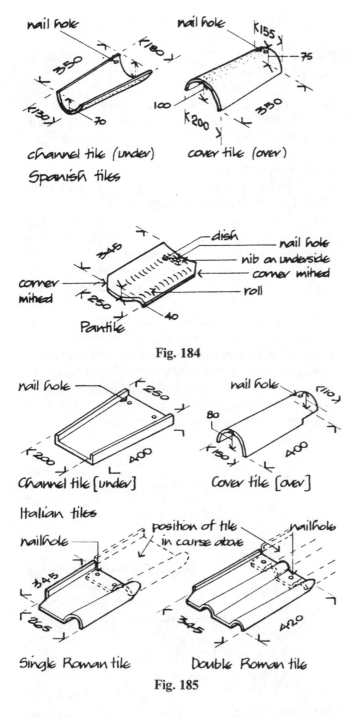

channel tile (under)

Spanish tiles

cover tile (over)

Fig. 184

Pantile

Channel tile [under]

Cover tile [over]

Italian tiles

Single Roman tile

Double Roman tile

Fig. 185

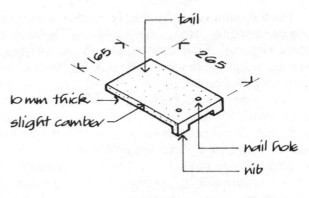

Fig. 186

Plain tiles are flat, rectangular roofing units size 265 by 165 with holes for nailing and nibs for hanging to battens as illustrated in Fig. 186. These small tiles are laid double lap down the slope of the roof because water running between the open joints between adjacent tiles runs on to the back of a tile double lapped under the joint. A plain tile roof covering is generally heavier than a comparable single lap tile roof. Plain tiles are less used than they were.

Concrete roofing tiles are very extensively used today as a substitute for good quality clay tiles. The tiles are manufactured from a mixture of carefully graded sand, Portland cement and water which is compressed in a mould. A thin top dressing of sand, cement and colouring matter is then pressed into the top surface of the tile. The tiles are then left under cover for some days to allow them to harden.

Concrete tiles are uniform in shape, texture and colour. The colours in which these tiles have so far been produced are very poor when compared with those of natural burned clay tiles. The colouring of these tiles is only in the top surface and in the course of time the colour tends to become washed out, leaving anaemic looking tiles. Concrete tiles are hard and dense and should be laid at slopes of from $17\frac{1}{2}$ to 35 degrees to the horizontal. A variety of shapes and sizes is manufactured as pantiles, flat interlocking and Roman tiles.

Concrete tiles are heavier than similar clay tiles.

Clay roofing tiles: Clay can be excavated, moulded and burned without any expensive or elaborate machinery and for centuries clay roofing tiles have been used in this country.

In general two qualities of tile are manufactured (a) hand-made tiles, and (b) machine-pressed tiles.

Hand-made tiles are made from a sandy clay which is pressed into shape by hand pressure so that even if some form of screw press is used the clay is not very heavily compacted. During burning the tiles shrink and there are quite noticeable variations which give a tiled roof so attractive an appearance. The tiles vary in colour from light brick red to almost black and many of the more expensive tiles are sand faced to produce the rough texture which is popular.

Hand-made tiles fairly readily absorb water and should not be laid at a pitch of less than 45 degrees to the horizontal. If laid at a less pitch these tiles may become saturated in winter and frost will then cause them to disintegrate.

Machine-pressed tiles are made from selected pure clays which are thoroughly ground to a fine condition. The clay is mixed with very little water and is heavily machine-pressed into tiles. Because little water is used to make the clay sufficiently plastic for moulding, the tiles do not shrink noticeably during burning and because the tiles are machine moulded they have smooth faces. These tiles are very hard and dense and do not absorb water readily and can be laid on a roof with a lower pitch than can the hand-made tiles. Machine-pressed tiles have successfully been laid at a pitch of as little as 30 degrees to the horizontal but the minimum pitch for these tiles is generally accepted as being 35 degrees to the horizontal. Because of their uniformity of shape, texture and colour, machine-pressed tiles do not make for so attractive a roof as the hand-made variety. Some machine-pressed tiles are faced with sand before being burned to give them a more attractive appearance. If clay roofing tiles have been made from clay which is free from lime and stones and the tiles have been hard burned and the tiles are laid at a suitable slope they will have a useful life of up to a hundred years or more. Good quality clay tiles weather well, which means they are resistant to damage by rain, frost, heat and all the dilute acids in industrial atmospheres.

Over the course of years the colour of clay tiles becomes somewhat darker and it is generally accepted that the appearance of a tile improves with age.

Single lap tiling

The types of single lap tiling used in the United Kingdom are:

(a) Interlocking tiles.
(b) Pantiles in which the under and over of Spanish tiles is incorporated in one tile.
(c) Single and double Roman tiles in which the unders and overs of Italian tiles are incorporated in one tile.
(d) Spanish tiles consisting of pairs of rounded unders and overs.
(e) Italian tiles consisting of flats unders and rounded overs.

Single lap tiles with ordinary side lap are laid at a pitch of not less than 30 degrees and concrete tiles with interlocking side lap at a pitch of from 25 degrees to 15 degrees.

Interlocking single lap tiles: A range of interlocking concrete and clay single lap tiles is manufactured. The concrete tiles are made from the same materials and in a similar manner to concrete plain tiles. They differ from ordinary single lap tiles in that one or more grooves in the vertical edges of the tiles, Fig. 188, interlock when the tiles are laid. The advantage of this side lock is that it does more effectively exclude wind and rain than a simple overlap. Concrete interlocking side lap tiles are made in various colours such as red, green, yellow and grey.

Clay interlocking pantiles are made from selected pure clays which can be moulded and burned without any appreciable loss of shape. In consequence a sophisticated system of grooves at the sides and head of the tiles interlock with matching grooves on adjacent tiles, to provide a very efficient barrier to penetration of wind and rain. These tiles are made in a variety of colours such as green brown and red and also with various glazed colour finishes.

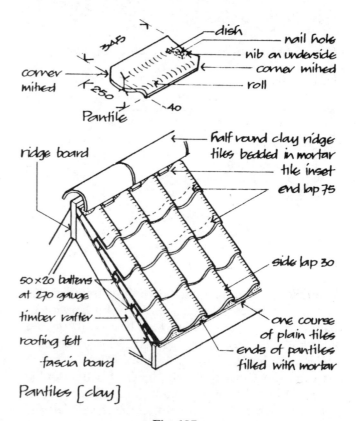

Fig. 187

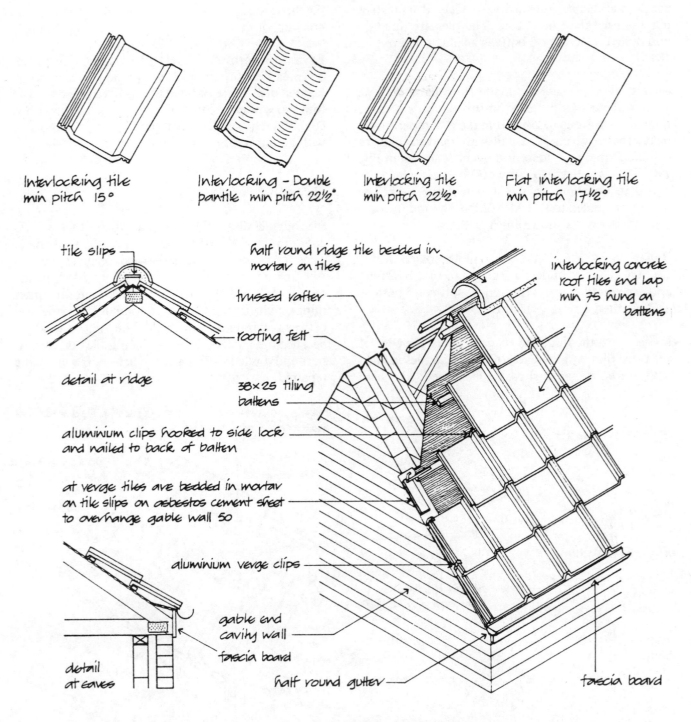

Interlocking tile
min pitch 15°

Interlocking - Double
pantile min pitch 22½°

Interlocking tile
min pitch 22½°

Flat interlocking tile
min pitch 17½°

tile slips

detail at ridge

aluminium clips hooked to side lock
and nailed to back of batten

at verge tiles are bedded in mortar
on tile slips on asbestos cement sheet
to overhange gable wall 50

detail
at eaves

half round ridge tile bedded in
mortar on tiles

trussed rafter

roofing felt

38×25 tiling
battens

aluminium verge clips

gable end
cavity wall

fascia board

half round gutter

interlocking concrete
roof tiles end lap
min 75 hung on
battens

fascia board

Fig. 188

Pantiles: Concrete pantiles are used extensively in the United Kingdom. A typical concrete pantile is illustrated in Fig. 188. Clay pantiles are hung on and nailed to softwood battens as illustrated in Fig. 187. The purpose of the mitred corners of these tiles is to facilitate fixing. But for the mitred corners there would be four thicknesses of tile at the junction of horizontal and vertical joints which would make it impossible to bed the tiles properly.

Single and double Roman tiles: Clay Roman tiles are less used than they were. The tiles are hung on and nailed to softwood battens as illustrated in Fig. 189.

Spanish tiles: These are made from natural clays. Few of these tiles are used in this country because they are laborious to fix and make a comparatively heavy roof covering. These tiles are made as pairs of rounded tapering unders and overs as shown in Fig. 190. They are fixed to softwood battens which are nailed to boarding up the slopes of the roof. The unders are nailed to the sides of battens and overs to top of battens, as shown in Fig. 190.

Italian tiles: These clay tiles are little used in this country. The flat unders are nailed to roof boarding and the rounded overs to vertical softwood battens as illustrated in Fig. 191.

Ridges: At the ridge the space between the edge of the ridge tiles and the back of single lap tiles is filled with mortar in which slips of tile are bedded. These tile slips are cut from plain tiles and serve to minimise cracking of the comparatively thick mortar filling.

Eaves: To provide a level bed on which mortar can be spread and onto which the single lap tiles can be bedded one course of plain tiles is fixed at eaves. The four types of clay single lap tiles are made in various colours such as red, brown, purple and grey and also with glazed finishes in certain colours.

Plain tiles

Size of plain tiles: The standard size of these tiles is 265 long by 165 wide and not less than 10 thick. These rather peculiar sizes derive, it seems, from 'An Act for the making of Tile' passed in 1477, in the reign of Edward IV. The Act required, among other things, that the tile be $10\frac{1}{2}$ in long, $6\frac{1}{4}$ in wide and between $\frac{1}{2}$ in and $\frac{1}{8}$ in thick. The Act was passed because tile makers were producing tiles of various sizes and this led to difficulties when roofs were being repaired.

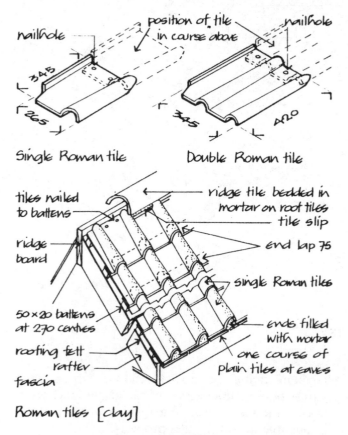

Fig. 189

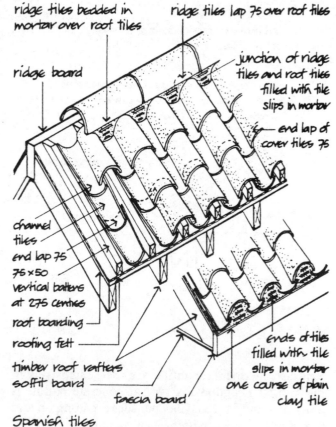

Fig. 190

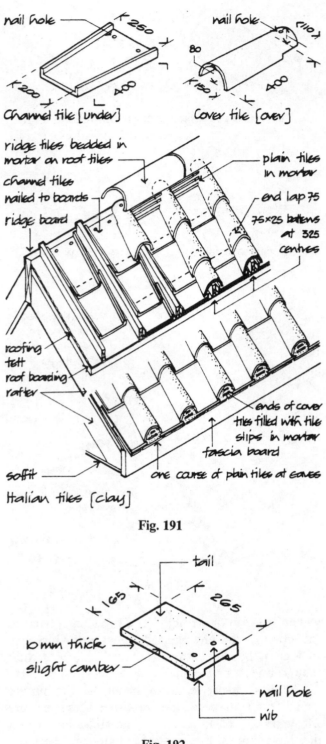

Fig. 191

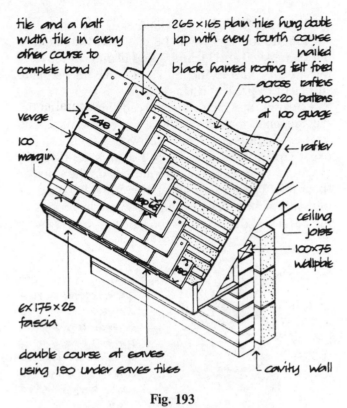

Fig. 193

Fig. 192

Nibs: Standard plain clay tiles today have projecting nibs at one end of the tile or one continuous nib. The tiles are secured to the sloping surface of the roof by hanging them by these nibs to horizontally fixed timber battens. A view of a plain tile is shown in Fig. 192.

Camber: The tiles are not perfectly flat but have a slight rise or camber in the back of them. The purpose of this camber is to prevent water being drawn up between the tiles by capillary action, as it would be if the tiles were absolutely flat.

Gauge and lap: Plain tiles are hung on 38×19 or 50×25 sawn softwood battens which are nailed across the rafters of the roof. The tiles are hung so that at every point on the roof there are at least two thicknesses of tile. This is illustrated in the view of part of a tiled roof shown in Fig. 193.

It will be seen that the tiles are hung so that their sides butt together and these joints are bonded up the slope of the roof. The tiles in every fourth course are nailed to the battens. This is a precaution to ensure that in high wind the tiles are not lifted off the roof. In very exposed positions every tile should be nailed to the battens. Each tile lies over the tile below it and also some 65 over the head of the tile below that. It is obvious that tiles must at least lie over half the length of the tile below to prevent water running between the butt side joints into the roof below. Plain tiles are laid double lap to prevent rain running off the back of one tile, in the joint between the two tiles below and then spreading out between tiles so that it runs into the roof. Figure 194 shows a few tiles laid without double lap. It will be seen that water creeps between tiles B and D, and C and D and runs in over the head of tiles E and F. The lap has to be made sufficient to prevent this.

The angle at which water spreads out between the tiles depends mainly on the slope of the tiles. The steeper the slope the less water will spread out. The usual lap for plain tiles is 65. For this lap tiles 265 long have to be hung at 100 intervals up the slope of the roof. The battens must therefore be fixed at intervals of 100. The distance between the centres of the battens is described as the gauge.

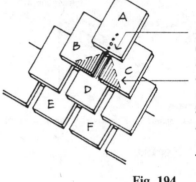

water running off back of tile A runs between B and C and spreads between tiles B and D and C and D as shown by hatched area and runs into roof over head of tiles E and F.

Fig. 194

Eaves: So that there are two thicknesses of tile at the eaves a course of eaves tile is used. These tiles are 190 long and 165 wide and are the first course of tiles to be hung as illustrated in Fig. 193. The tails of the eaves tiles and the tile course above overhang the fascia board some 40 to shed water into the eaves gutter.

Ridge: Any one of four standard sections of clay ridge tile may be used to cover the ridge. The four standard sections of clay ridge tile are illustrated in Fig. 195. Tile courses are hung up to the ridge board and a special top course of tiles 190 long is used. The ridge tiles, whichever of the four sections is used, have their edges bedded in fillets of cement mortar spread on the back of the top course tiles. The ridge tiles butt together and so that the joint can be satisfactorily filled with mortar the ends of all ridge tiles are solidly filled with mortar. This prevents the mortar in the joints from falling in. A section through a tile ridge is illustrated in Fig. 196.

Hips: The hips can be covered with ridge tiles bedded in exactly the same way that they are on ridges. To prevent the tiles slipping down the hip a

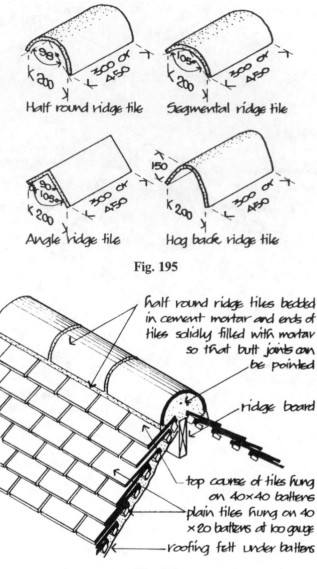

Half round ridge tile

Segmental ridge tile

Angle ridge tile

Hog back ridge tile

Fig. 195

half round ridge tiles bedded in cement mortar and ends of tiles solidly filled with mortar so that butt joints can be pointed

ridge board

top course of tiles hung on 40×40 battens

plain tiles hung on 40 ×20 battens at 100 gauge

roofing felt under battens

Fig. 196

galvanised iron or wrought-iron hip iron is fixed to the hip or fascia. The plain tiles next to the hip have to be cut to fit against the side of the hip rafter so that they lie under the hip tiles.

The tiles which are cut to the side of the hip are usually cut from special tile-and-a-half tiles which are 248 wide by 265 long. These tile-and-a-half width tiles have three nibs so that after being cut there is a nib to hang them on. A hip covered with ridge tiles has a poor appearance.

Hip tiles are manufactured which can be nailed to the hip rafter so that they lap one over the other and the tail of these tiles lines up with the courses of plain tiles on the roof. The two standard sections of hip tiles are illustrated in Fig. 197. Each tile is nailed to

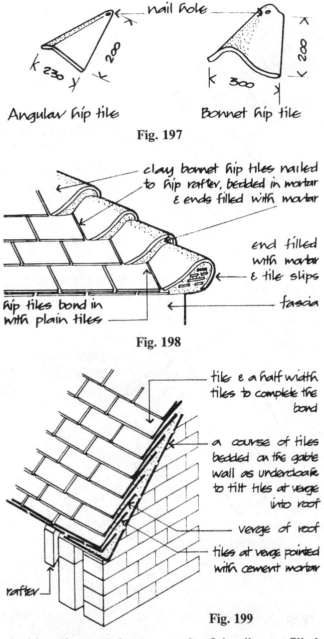

Angular hip tile · Bonnet hip tile

Fig. 197

clay bonnet hip tiles nailed
to hip rafter, bedded in mortar
& ends filled with mortar

end filled
with mortar
& tile slips

fascia

hip tiles bond in
with plain tiles

Fig. 198

tile & a half width
tiles to complete the
bond

a course of tiles
bedded on the gable
wall as undercloak
to tilt tiles at verge
into roof

verge of roof

tiles at verge pointed
with cement mortar

rafter

Fig. 199

the hip rafter and the open ends of the tiles are filled with cement mortar. Of the two sections the bonnet hip gives the most pleasing finish to a roof as the curve of its tail carries the line of the tiles smoothly over the hip as illustrated in Fig. 198.

Verge: The verge of plain tiling at a gable end of a roof is usually bedded on the gable end wall and tilted up slightly to prevent water running off the tiles down the gable end wall. Tile-and-a-half width tiles are used in every other course to complete the bond. The construction of a tiled verge is illustrated in Fig. 199.

Slates

Roofing slates: The traditional materials that were used for roof slating are Welsh slates and Westmoreland slates. The majority of slated roofs were covered with Welsh slates. Today very few natural slates are quarried in Wales and Westmoreland. Repairs to existing slated roofs are carried out with used slates and reslating of old roofs and slating new roofs is generally executed with fibre cement slates. Fibre cement slates are manufactured from fibres, cement and water in sizes similar to natural slates.

Fibre cement slates were originally made from cement and water reinforced with asbestos fibres. Because of the hazard to the health of operatives engaged in the manufacture of asbestos fibre-based materials, the original asbestos fibre reinforcement has been replaced with natural and synthetic fibres. These man-made slates are made from pigmented Portland cement and water, reinforced with natural and synthetic fibres. The wet mix is compressed to slates which are cured to control the set and hardening of the material. The standard slate, known by the trade name 'Eternit', is 4 thick, with a matt coating of acrylic finish and a protective seal on the underside against efflorescence and algae growth. The standard slate is rectangular and holed for nail and rivet fixing. These slates are made in three colours blue/black, brown and rose and with either square tails or curved, angled or chamfered tails for decoration.

The standard fibre cement slate is uniform in shape, colour and texture. The current cost of covering a roof with standard fibre cement slates is about half that of covering the same roof with natural slates. Fibre cement slates have a useful life of about 30 years in normal circumstances as compared to a useful life of 100 years or more for sound natural slates.

Standard fibre cement slates are made in sizes of 600×300, 500×250 and 400×200 as illustrated in Fig. 200, with a comprehensive range of fittings for ridge, eaves, verges and valleys.

Two somewhat more expensive forms of fibre cement slates are made to simulate the appearance of natural slate in colour and texture and the other with a random brick coloured surface.

Fixing: The slates are laid at a pitch (slope) of from 20 to 45 degrees, the longer slates being at the lower

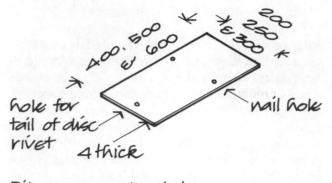

hole for
tail of disc
rivet

nail hole

4 thick

Fibre cement slate

Fig. 200

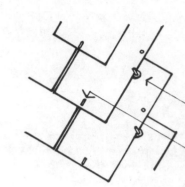

copper disc rivets
fit between slates
and tail of rivet is
bent down to hold
tail of slate above

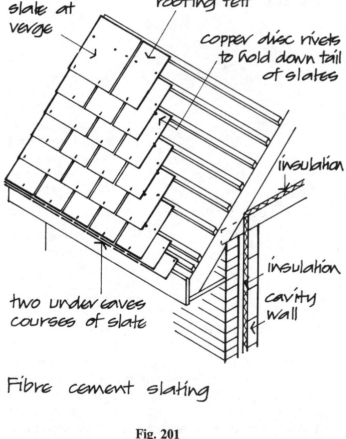

slate and a
half width
slate at
verge

500×250 fibre cement
slates, centre nailed to
38×25 battens at 210
gauge on reinforced
roofing felt

copper disc rivets
to hold down tail
of slates

insulation

insulation

cavity
wall

two undereaves
courses of slate

Fibre cement slating

Fig. 201

pitch, double lap down the slope with a minimum end lap of from 70 to 110 depending on the length of slate, pitch and exposure. The slates are fixed to 38 × 19 or 38 × 25 battens nailed to rafters over roofing felt at a gauge calculated as

$$gauge = \frac{length\ of\ slate - lap}{2}$$

Each slate is centre nailed to battens with two copper nails as illustrated in Fig. 201. Because these slates are thin, comparatively lightweight and somewhat brittle they are secured with copper disc rivets that fit between slates in the undercourse and are bent down over the tail of the slate above to hold it down as illustrated in Fig. 201. The purpose of these disc rivets is to hold down the tail of each slate against wind uplift that might otherwise cause the slate to fracture.

Eaves: So that there will be two thicknesses of slate at eaves, special length undereaves slates are used and head nailed to battens as illustrated in Fig. 201.

Ridge: Special lengths of slate are used for the top courses at ridge to provide two thicknesses of slate cover as illustrated in Fig. 202. The ridge is covered with one of the angled or curved fibre cement ridge fittings that overlap and are secured with screws or clips as illustrated in Fig. 202.

Hips: At hips, slates are mitre cut to form a close mitre finish by using slate and a half width slate to bond in with the rest of the covering. The mitred hip is weathered with Code 3 lead soakers that are hung over the head of slates.

Verge: To complete the bond of slates at verges with whole slates, a slate and a half width slate is used in every other course as illustrated in Fig. 201.

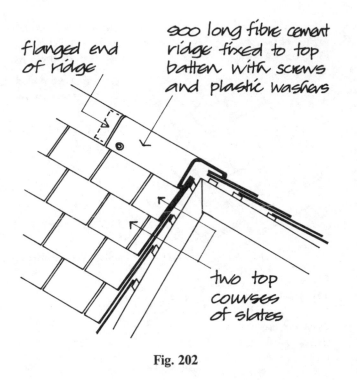

flanged end of ridge

900 long fibre cement ridge fixed to top batten with screws and plastic washers

two top courses of slates

Fig. 202

Natural slates

Welsh slates were extensively quarried in the mountains of Wales. The stone from which they are split is hard and dense and varies in colour from light grey to purple. In the quarry the stone consists of thin layers of laminae of hard slate with a very thin somewhat softer layer of slate between the layers. By driving a wedge of steel into the stone between the layers it can be quickly split into fairly large quite thin slates of thicknesses varying from 4 to 10. The splitting results in slates of varying thickness and the thicker slates are classified as best, strong best, and medium, and the thinner slates as seconds and thirds. There is no agreed minimum thickness for the above classification which therefore seems somewhat pointless. The slates are cut to a variety of sizes varying from 610 long by 355 wide to 255 × 150. The most commonly used sizes are 610 × 355, 510 × 255, 460 × 230 and 405 × 205. For many years each size of slate was given a particular name, such as Countess for the 510 × 255 slates. No useful purpose is served in giving a name to each specific size of slate and it is hoped that before long naming of slate sizes will be discontinued.

Good quality Welsh slates are hard and dense and do not readily absorb water. They are not affected by frost or the dilute acids in industrial atmospheres and will have a useful life as a roof covering of very many years. The durability of a slate depends upon its density and the most reliable guide to this is the amount of water it will absorb if immersed in water for a set time. Slates which absorb the least water will have a useful life as a roof covering of a hundred years or more, and can be laid at a slope of as little as 25 degrees to the horizontal.

Westmorland slates are quarried from stone in the mountain region of the Lake District. The stone varies in colour from blue to green with flecks and streaks of browns and greys. The stone is very hard and dense, and consists of irregular layers or laminae of hard stone, separated by very slightly softer stone. The laminae do not run in regular flat planes as in Welsh slate and the stone is more difficult to split than the Welsh. As a consequence it is not so economic to split slates and cut them to uniform size and Westmorland slates are commonly used in random sizes. These slates are split to thicknesses of from 7 to 19 and the slates have rough surfaces and sharp irregular edges.

The best quality Westmorland slates are very hard and so dense that they absorb practically no water no matter how long they are immersed in water. These slates are practically indestructible. Because of the considerable labour required to cut the stone Westmorland slates are expensive and generally only used for more expensive building works.

Fixing Welsh slates: Slates are fixed to 50 × 25 sawn softwood battens by means of copper composition nails driven through holes which are punched in the head of each slate. Galvanised iron or iron nails should not be used as they will in time rust and allow the slates to slip out of position. Two holes are punched in each slate some 25 from the head of the slate and about 40 in from the side of the slate as illustrated in Fig. 203. The battens are nailed across the roof rafters and the slates nailed to them so that at every point on the roof there are at least two thicknesses of slate and so that the tail of each slate laps 75 over the head of the slate two courses below. This is similar to the double lap arrangement of plain tiles and is done for the same reason so that slates are double lapped. Because the length of slates vary and the lap is usually constant it is necessary to calculate the spacing or gauge of the battens. The formula for this calculation is:

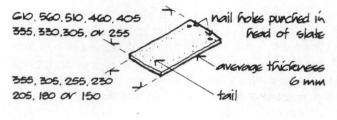

610. 560. 510. 460. 405
355. 330. 305. or 255

355. 305. 255. 230
205. 180 or 150

nail holes punched in head of slate

average thickness 6 mm

tail

Fig. 203

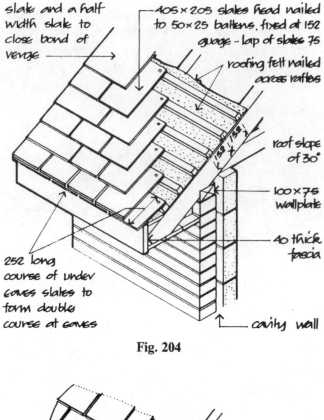

slate and a half width slate to close bond of verge

405 × 205 slates head nailed to 50 × 25 battens, fixed at 152 gauge - lap of slates 75

roofing felt nailed across rafters

153 × 38

roof slope of 30°

100 × 75 wallplate

40 thick fascia

cavity wall

252 long course of under eaves slates to form double course at eaves

Fig. 204

$$\text{gauge} = \frac{\text{length of slate} - (\text{lap} + 25)}{2}$$

For example the gauge of the spacing of the battens for 510 × 255 slates is:

$$\frac{510 - (75 + 25)}{2} = 205$$

Figure 204 illustrates the fixing of slates to a pitched roof.

Roofing felt should be nailed across the rafters under the battens to keep wind out of the roof.

Head nailed and centre nailed slates: Slates are usually fixed by means of nails driven through holes in the head of slates. This is the best method of fixing slates as the nail holes are covered with two thicknesses of slate so that even if one slate cracks, water will not get in. But if long slates such as 610 × 305 are head nailed on a shallow slope of say 30 degrees or less it is possible that in high wind the slates may be lifted so much that they snap off at the nail holes. In exposed positions on low pitch roofs it is common to fix the slates by centre nailing them to battens. The nails are not driven through holes exactly in the centre of the length of the slate but at a distance equal to the gauge down from head of slate, so that the slate can double lap at tails as illustrated in Fig. 205. It will be seen that with this method of fixing there is only one thickness of slate over each nail hole so that if that slate cracks water can get into the roof.

When roof boarding is used to keep the roof space wind tight the slates, whether head or centre nailed, are usually nailed directly to the boarding through the roofing felt as illustrated in Fig. 206.

Eaves: So that there shall be two thicknesses of slates at the eaves a course of undereaves slate is used. These slates are cut to a length equal to the gauge + lap + 25 of the slating as illustrated in Fig. 204.

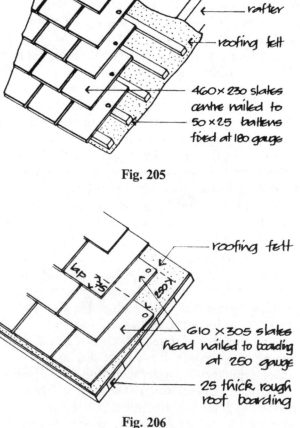

rafter

roofing felt

460 × 230 slates centre nailed to 50 × 25 battens fixed at 180 gauge

Fig. 205

roofing felt

lap 75

250

610 × 305 slates head nailed to boarding at 250 gauge

25 thick rough roof boarding

Fig. 206

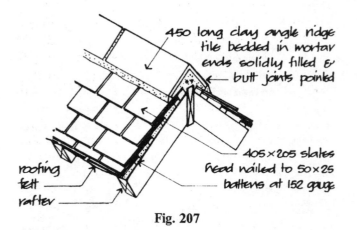

Fig. 207

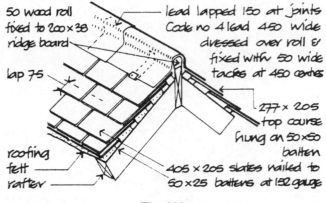

Fig. 208

Ridge: The cheapest way of covering the ridge of slated roofs is to use one of the clay ridge tiles which are bedded and pointed as they are on the ridge of a tiled roof. The top course of slates is usually shorter than the ordinary slates, being gauge + lap + 50 or 75 in length, as illustrated in Fig. 207. Another method of covering a slate ridge is to finish it with strips of lead sheet dressed over a wood roll and secured with lead clips as shown in Fig. 208.

Hips: These are covered with either ridge tiles or lead sheet in the same way that the ridge is covered.

Verge: At the verge of a slate slope, slate-and-a-half width slates are used to complete the bond and the slates are tilted up slightly as are tiles at the verge of a tiled roof.

Sheet metal coverings to pitched roofs

For many years during the middle of this century it was practice to construct low pitched roofs for houses, schools and other buildings. A pitched roof is generally defined as one with slopes of 10 degrees or more to the horizontal and a low pitched roof has slopes of from 10 degrees to 30 degrees to the horizontal. The disadvantages of a flat roof were known and a steeply pitched roof is not always an attractive feature of small buildings. For example a small bungalow with a pitched roof covered with tiles does not usually look attractive as the great area of tiles dominates the lesser area of wall below. A low pitched roof is a happy compromise between flat and steeply pitched roofs. It at once looks attractive, gives reasonable insulation against loss of heat and provides roof space in which water storage tanks can be housed.

The principal coverings for low pitched roofs are copper and aluminium strips, both of which are comparatively light in weight and therefore do not require heavy timbers to support them and both have a useful life as a roof covering of many years. The roofs are constructed as single slope roofs or as low pitched roofs with timber rafters pitched to a central ridge board with or without hipped ends. The construction of the rafters and their support with purlins and struts, or purlins and light timber trusses is similar to that for other pitched roofs as previously explained.

Copper or aluminium strips of 450 or 600 width and up to 8.0 long are used to cover these low pitched roofs. No drips or double lock cross welts or other joints transverse to the fall are used with strips up to 8.0 long and because of this the labour in jointing the sheets is less than that required with batten or conical roll systems of covering and consequently copper or aluminium strip coverings are comparatively cheap. Because of the great length of each strip the fixing cleats used to hold the metal strips in position have to be designed to allow the metal to contract and expand freely.

Standing seams: The strips of metal are jointed down the slope of the roof by means of a standing seam joint which is a form of double welt and is left standing up from the roof as shown in Fig. 209, which is a view of part of a low pitched roof. The completed standing seam is constructed so that there is a gap of some 13 at its base which allows the metal to expand without restraint and this is illustrated in Fig. 209. The lightweight metal strips have to be secured to the roof surface at intervals of 300 along the length of the standing seams. This close spacing

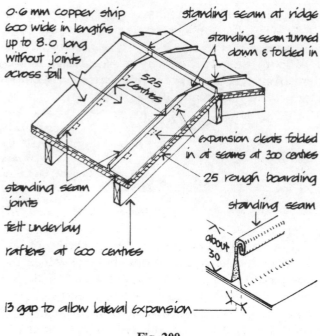

Fig. 209

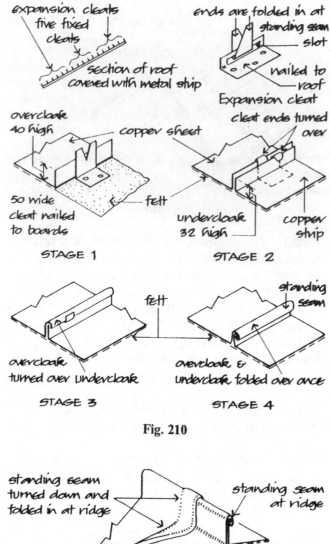

Fig. 210

Fig. 211

of the fixing cleats is necessary to prevent the metal drumming in windy weather.

Two types of cleats are used, fixed cleats and expansion cleats. Five fixed cleats are fixed in the centre of the length of each strip and the rest of the cleats are expansion cleats. Figure 210 illustrates the arrangement of these cleats. The fixed cleats are nailed to the roof boarding through the felt underlay and Fig. 210 illustrates the formation of a standing seam and shows how the fixed cleat is folded in.

The expansion cleats are made of two pieces of copper strip folded together so that one part can be nailed to the roof and the second piece, which is folded in at the standing seam, can move inside the fixed piece. Figure 210 illustrates one type of expansion cleat used.

Ridge: The ridge is usually finished with a standing seam joint as illustrated in Fig. 209, but as an alternative a batten roll or conical roll may be used. Whichever joint is used at the ridge the standing seams on the slopes of the roof have to be turned down so that they can be folded in at the ridge. This is illustrated in Fig. 211.

Where the slope of the roof finishes at the parapet or wall the strips of metal are turned up as an upstand and finished with an apron flashing. This is illustrated in Fig. 212 which also illustrates the cutting and turning down of the standing seam.

Eaves: Because copper or aluminium are generally considered to be attractive coverings to roofs, the roof is not hidden behind a parapet wall, and the roof slopes discharge to an eaves gutter as illustrated in Fig. 213.

Verge: The verge of low pitched roofs can be finished with batten roll or conical roll as illustrated for flat roof coverings.

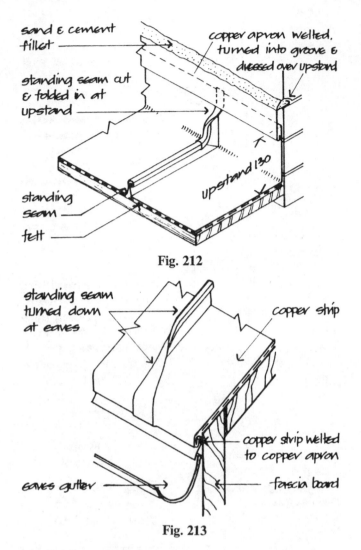

sand & cement fillet

copper apron welted, turned into groove & dressed over upstand

standing seam cut & folded in at upstand

upstand 130

standing seam

felt

Fig. 212

standing seam turned down at eaves

copper strip

copper strip welted to copper apron

eaves gutter

fascia board

Fig. 213

Felt underlay: It is of importance that strip metal coverings be laid on an underlay of bitumen-impregnated roofing felt laid across the roof boards and nailed to the boards with butt side joints. The felt allows the metal strips to expand and contract without restraint.

Resistance to fire

The space inside a pitched roof is a void space that should be separated from other void spaces or cavities by cavity barriers that seal the junction of the cavities to prevent unseen spread of smoke and flames. The cavity in a wall is generally separated from the cavity in a pitched roof by the cavity barrier at the top of the cavity in the wall.

Separating walls, party walls, between semi-detached and terraced houses should resist the spread of fire from one house to the other by being raised above the level of the roof covering at least 375 or by being built up to the underside of the roof covering so that there is a continuous fire break.

The traditional roof coverings slate, tile and non-ferrous sheet metal do not encourage spread of flame across their surface and are, therefore, not limited in use in relation to spread of fire between houses.

Resistance to the passage of heat

The 1989 revision of the requirements for the insulation of roofs is a maximum U value of 0.25. The thermal resistance of a pitched roof of conventional construction is small so that it is necessary to form a layer of insulating material to bring resistance to the passage of heat up to the required value.

There are two methods of determining the thickness of insulation required to provide a maximum U value of 0.25. The first method is to relate the thickness of insulation required to the thermal conductivity of the insulating material to be used, ignoring the thermal resistance of the roof itself. This produces a reasonably accurate indication of insulation thickness because the thermal resistance of the roof is small in comparison to that of the insulating layer. For example, if an insulant with a thermal conductivity of 0.04 is selected the required thickness will be 145.

The second method is to make a calculation of the thermal resistance of the roof and then determine the required thickness of insulation. Assume a pitched roof covered with 10 thick tiles on felt underlay with 13 plasterboard ceiling.

	Thermal resistance (m^2K/W)
External surface	0.04
Tiles	0.01
Loft space	0.18
Plasterboard	0.03
Inside surface	0.10
Total	0.36

Thermal resistance required is

$$\frac{1}{0.25} = 4.0 \ m^2K/W$$

Additional resistance required
$$= 4.0 - 0.36$$
$$= 3.64 \ m^2K/W$$

Thickness of insulation required
$$= 3.64 \times 0.04 \times 1000$$
$$= 145.6 \ mm$$

The thickness of insulation can be varied by using an insulant of a lower or higher thermal conductivity. Using an insulant with a thermal conductivity of 0.02 the thickness required is 73. As there is no advantage in using a thin insulation material inside a roof space, considerations of cost and convenience in fixing or laying the material are usually determining factors.

The most convenient and economical place to fix insulation in a pitched roof is between or across the top of the ceiling joists. With insulation at ceiling joist level the roof above is a cold roof. The materials most used in this situation are mineral wool mats or rolls of fibre glass or rockwool spread across or between the joists or loose fill spread between the joists on top of the ceiling finish as illustrated in Fig. 214.

For effective insulation, the insulating layer should extend right across the roof in both directions and be joined to or overlap insulation to walls and should extend up to and be continued over any roof hatch fixed in ceilings for access to roofs as illustrated in Fig. 215.

With mineral wool mats or rolls and loose fill insulation spread between the ceiling joists there will be to an extent, thermal or cold bridges across the timber ceiling joists, which have less thermal resistance than most insulating materials.

With mats or rolls of mineral fibre spread across joists there is a possibility of the loose material being compressed and losing efficiency as an insulator, under walkways for access in roof spaces. Boarded access ways inside roofs should, therefore, be raised on battens above the level of the insulation. The insulants commonly used for cold pitched roofs are detailed in table 16.

Table 16. Insulation Materials

Pitched roof cold roof	Thickness	U valve W/m²K
Glass fibre		
rolls laid across or between ceiling joists	60, 80, 100, 150, 200	0.04
semi-rigid batts laid across or between ceiling joists	80, 90, 100, 120, 140, 180, 200	0.04
Rockwool		
rolls laid across or between ceiling joists	80, 100, 150	0.037
granulated fibre spread between ceiling joists		0.043

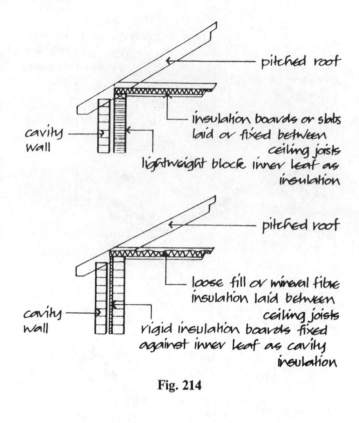

Fig. 214

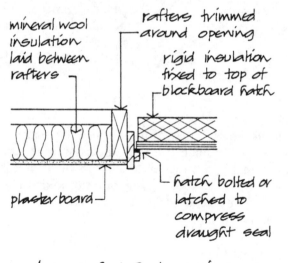

Access hatch to roof space

Fig. 215

Condensation–ventilation

The vapour pressure from warm, moist air in rooms below the ceiling of roofs may cause moisture vapour to penetrate the ceiling and insulation. This will be most likely above bathrooms in which there will be the heaviest concentration of warm moist air. Rooms such as bathrooms and kitchens should be provided with means of natural or mechanical extract ventilation to reduce the build up of moisture vapour in air. The 1989 revision of Approved Document F, giving practical guidance to meeting the requirement in The Building Regulations 1985 for ventilation to dwellings, provides a natural extract duct with a diameter of 150 and ventilation openings of at least one-twentieth of the floor area of kitchens, a duct with a diameter of 150 for bathrooms or mechanical extract ventilation for occasional rapid ventilation capable of extracting at a rate of not less than 10 air changes per hour for both bathrooms and kitchens.

Any moisture vapour that finds its way through a ceiling and insulation will condense to water on the cold side of the insulation where it may saturate insulation and reduce its efficiency as an insulant and may also saturate timber and encourage decay.

There are two methods of reducing the likelihood of condensation of moisture vapour on the cold side of insulation. Both are often used to check condensation. The first is to fix a vapour check across the warm side of the insulation and the second to ventilate the roof space above the insulation, to outside air. A vapour check is a continuous layer of some material which is impermeable to moisture vapour, such as polythene sheet, which is spread or fixed below the insulation. Sheets of polythene 250 gauge are used with the edges of sheets overlapped and continued to overlap any vapour check to walls. It is impractical to form a continuous barrier particularly at junctions of walls and ceilings and around pipes penetrating the ceiling.

With mineral wool or loose granule insulation spread over a ceiling there are practical difficulties in spreading a vapour check below them either over or between joists, without damage to the vapour check and because of electrical cables and water service pipes penetrating the ceiling. The most practical way of fixing a vapour check is to use one of the plasterboards that have a vapour check film bonded to the back of the boards. With these plasterboards there is no effective way of sealing the edges of boards with the film.

Organic closed cell insulants in the form of boards, such as extruded polystyrene, are substantially impermeable to water vapour and will act as a vapour check. Where these boards are fixed across ceiling joists and closely side butted there is no need for a vapour check.

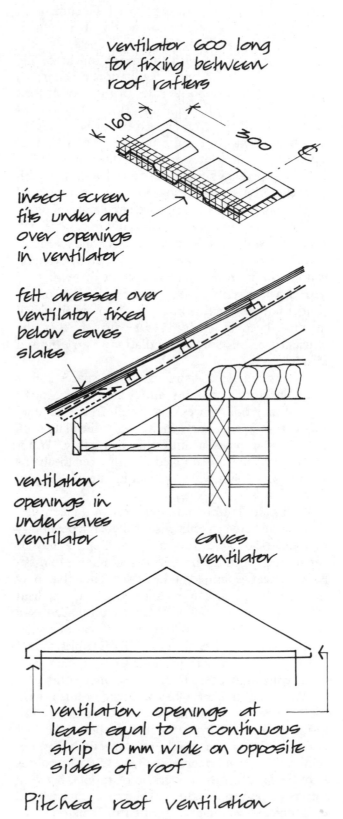

ventilator 600 long for fixing between roof rafters

160

300

insect screen fits under and over openings in ventilator

felt dressed over ventilator fixed below eaves slates

ventilation openings in under eaves ventilator

eaves ventilator

ventilation openings at least equal to a continuous strip 10 mm wide on opposite sides of roof

Pitched roof ventilation

Fig. 216

The 1989 revision of Approved Document F, giving practical guidance to meeting the requirements of The Building Regulations 1985 for the ventilation of roofs with the insulation at ceiling level, cold roofs, requires ventilation openings to pitched roofs on opposite sides of the roof at least equal to a strip 10 wide all along the roof as illustrated in Fig. 216. These ventilation openings may be formed in the soffit of projecting eaves or at eaves level under slates or tiles as illustrated in Fig. 216. The openings should be covered with fine mesh to exclude bees and wasps. In addition ventilated ridges may be fitted to improve circulation of air.

The movement of air, which is unpredictable, will vary from still, cold weather to gusty windy weather and the necessary wind pressure needed to promote ventilation in roofs may not coincide with cold periods when ventilation is most required to control condensation. The fine insect mesh to openings will in time become clogged with dirt so that the efficiency of this sort of ventilation is very hit and miss.

A consequence of ventilating roof spaces is that water in storage cisterns and water service pipes located in roof spaces may well freeze in cold weather. It is necessary, therefore, to insulate such cisterns and pipes against this possibility. Water services may now be supplied directly from the mains supply (see Volume 5) so obviating the necessity to fix cisterns in roof spaces.

With mineral fibre mat and loose granule insulation spread between ceiling joists there is a possibility that electrical cables run in the roof space may be surrounded with the insulation and unable to dissipate the heat generated in the cables. The cables may then overheat and cause a short circuit or the heat may start a fire in combustible insulation. Cables in roof spaces should, therefore, be run clear of insulation or be of larger capacity and PVC cables should not be in contact with expanded polystyrene insulants which may cause them to become brittle.

Where the roof space below pitched roofs is used for storage or occasional use or there are rooms inside the roof the insulation is fixed below the roof covering. One of the organic insulants in the form of semi-rigid or rigid boards is often used. The boards may be fixed on the underside of rafters, between rafters or on top of rafters. The most practical and economical position to fix the boards is on top of the rafters where there will be least cutting to waste and fitting around intersecting timbers or between timbers. The advantage in using one of the closed cell organic insulants is that the material will serve as both insulation and as a vapour check to this form of warm roof. With insulation directly under the roof covering there is good reason not to ventilate the roof space.

Semi-rigid mineral wool boards or batts of fibre glass or rockwool can be friction-fitted between roof rafters and held in place with nails and a layer of polythene fixed to the underside of the rafters as a vapour check.

The most straightforward method of applying the insulation is to fix it across the top of rafters as a form of sarking, the name given to a material used to exclude wind from the roof space. Boards of one of the organic insulants such as XPS, PIR or PUR are nailed across rafters. Insulants such as XPS which are substantially impervious to moisture vapour, do not need a moisture vapour check beneath them. With insulation boards on top of rafters it is necessary to fix counter battens over roofing felt, under tiling or slating battens, so that any water that penetrates can run down to eaves.

Details of insulants suitable for warm roof insulation are given in table 17.

Table 17. Insulation Materials

Pitched roof warm roof	Thickness	U valve W/m²K
Glass fibre semi-rigid batts friction fit between rafters 1175 × 570 or 370	80, 90, 100, 120, 140, 180, 200	0.04
XPS boards fixed on top of rafters as sarking 2500 × 600	30, 40, 50, 60, 70, 80	0.025
PIR boards aluminium foil faced fixed on top of rafters as sarking 2400 × 1200	20, 25, 30, 35, 50	0.022
PUR boards glass tissue and aluminium foil faced fixed to top of rafters as sarking 1200 × 600 or 450	20, 25, 30, 35, 40, 50	0.022

XPS extruded polystyrene
PIR rigid polyisocyanurate
PUR rigid polyurethane

FLAT ROOFS

The traditional form of roof construction in this country has, for many centuries, been the timber framed pitched roof covered with slates or tiles. With the initial use of reinforced concrete during the early part of the twentieth century, a style of building was adopted using flat roofs that emphasised the thin horizontal elements of floors and roofs. This horizontal roof form, common to buildings in dry climates, was quite alien to the rainfall of the climate of northern Europe. As a revolt against the excesses of decoration that had persisted for some time, the simple and often elegant form of what is sometimes called the modern movement, was a welcome breath of fresh air. For some years it became fashionable to construct flat roofs in concrete and timber in the name of modern, often with unhappy results when the purity of form of the early examples of the modern movement were abandoned.

For about fifty years many buildings were constructed with flat roofs. It was accepted that the flat roof coverings, asphalt and felt, had a limited life and would have to be recovered every twenty years or so. There were some failures due to poor construction or detailing, but by and large there was no great alarm. It was with the adoption of increasingly higher standards of insulation for roofs, provoked by the increase in expectation of thermal comfort and increase in fuel prices, that flat roofs got a bad name. The consequence of the initial, ill-considered use of thermal insulating materials under flat roof coverings was that the dimensional changes that these coverings suffer between hot sunny days and following cold nights was greatly increased, which was a prime cause of failures. Tar and bitumen base products such as asphalt and felt are admirable waterproof membranes but quite unsuited to the extremes of temperature that a flat roof may suffer. There is nothing intrinsically wrong with an insulated flat roof structure providing the insulation is above roof membranes that will suffer deterioration from temperature extremes, as in the comparatively recent use of the inverted or upside down roof.

Strength and stability

Timber flat roof

The construction of a timber flat roof is similar to the construction of a timber upper floor. Softwood timber joists 38 to 75 thick and from 75 to 225 deep are placed on edge from 400 to 600 apart with the ends of the joists built into or on to or against brick walls and partitions.

Tables in Approved Document A, giving practical guidance to meeting the requirements for dwellings of up to three storeys for single families, give sizes of joists for flat roofs related to span and loads for roofs with access only for maintenance and repair and also for roofs not limited to access for repair and maintenance.

Strutting between joists: Solid or herringbone strutting should be fixed between the roof joists for the same reason and in like manner to that used for upper timber floors.

Roof deck: Boards which are left rough surfaced from the saw are the traditional material use to board timber flat roofs. This is called rough boarding and is usually 19 thick and cut with square, that is plain edges. Plain edged rough boarding was the cheapest obtainable and used for that reason. Because square edged boards often shrink and twist out of level as they dry, chipboard or plywood is mostly used today to provide a level roof deck. For best quality work tongued and grooved boards were often used.

End support of joists: If there is a parapet wall around the roof, the ends of the roof joists may be built into the inner skin of cavity walls or supported in metal hangers. The joists can bear on a timber or metal wall plate or be packed up on slate or tile slips as described for upper floors. The ends of the roof joists are sometimes carried on brick corbel courses, timber plate and corbel brackets or on hangers in precisely the same way that upper floor joists are supported. The ends of roof joists built into solid brick walls should be given some protection from dampness by treating them with a preservative.

Timber firring: Flat roofs should be constructed so that the surface has a slight slope or fall towards rainwater outlets. This slope could be achieved by fixing the joists to a slight slope but the ceiling below the roof would then also be sloping. It is usual to provide a sloping surface to the roof by means of firring pieces. These consist of either tapered lengths of fir (softwood) nailed to the top of each joist or varying depth lengths of softwood nailed across the joists as shown in Fig. 217. Tapered firring is used for

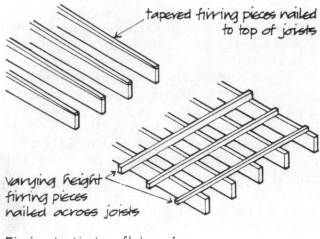

tapered firring pieces nailed to top of joists

varying height firring pieces nailed across joists

Firring to timber flat roof

Fig. 217

roofs covered with chipboard or wood wool slabs and the varying depth firring for boards laid parallel to the slope of the roof so that variations in the level of the boards do not impede the flow of rainwater down the shallow slope.

As an alternative to firring, some insulating boards are cut or made to a slight wedge section to provide the necessary fall to a roof.

Reinforced concrete roofs

All the types of reinforced concrete construction described for floors can equally well be used for roofs and the details of construction and advantages of each apply also to their use for flat roofs. The loads on roofs are usually somewhat less than those of floors and the thickness of a concrete roof will usually be less than that of a floor of similar span. The concrete of monolithic reinforced concrete roofs, and the constructional concrete topping of the other types of concrete roof, is usually finished off level. If the surface is to be laid to a slight slope or fall the concrete roof will be finished with a screed of cement and sand, as described for concrete floors, with the top surface of the screed finished to the falls required. The least thickness of the screed will be from 20 to 25.

A reinforced concrete roof will usually span the least width between the external walls or external walls and internal load bearing walls and will be supported on the walls and partitions in the same way as reinforced concrete floors.

A reinforced concrete roof provides poor insulation against loss or gain of heat, and some material which is a good thermal insulator should be incorporated in the construction of the roof or a lightweight concrete slab construction be used (see Volume 4). One way of improving the insulation of a concrete roof is to use a lightweight aggregate instead of sand for the screed which is spread over the concrete roof to provide a sloping surface for the roof covering. This screed can be made of lightweight aggregate in lieu of sand. The lightweight aggregates in common use are foamed slag, pumice and vermiculite. These three minerals are all porous and it is the air trapped in the minute pores of the materials which at once makes them light in weight and good thermal insulators.

Foamed slag is formed by spraying water on the molten slag which is poured off molten iron in blast-furnaces. The water causes the slag to expand into a porous lightweight mass. The slag is crushed into small particles. This is a moderately cheap material which is a good insulator. A screed 25 thick made of foamed slag and cement will appreciably improve the thermal insulation of a concrete roof.

Pumice is a rock of volcanic origin which is porous, lightweight and a good thermal insulator. It is crushed to small particles, mixed with cement and used as a screed. This is an expensive material not much used.

Vermiculite is a micaceous mineral which consists of fine layers of material closely packed. When it is heated the fine layers open out and gases are trapped in the many spaces between the expanded layers. After heating it is described as expanded or more usually 'exfoliated' vermiculite. The word exfoliated describes the opening out of the thin layers, or foliage, when the material is heated. A screed made of vermiculite and cement improves the insulation of a concrete flat roof.

The disadvantage of lightweight screeds, laid to falls is that it takes some considerable time for the water in the screed to dry out which necessitates the use of venting layers for felt and ventilators for asphalt coverings to allow moisture vapour to escape. With increase in the requirements for insulation the least thickness of screed required is quite considerable.

Resistance to weather

FLAT ROOF COVERINGS

Roofs are defined in relation to the angle of slope of their weather plane to the horizontal as:

Flat roof from 1° and up to 5°
Sloped roof over 5° and up to 10°
Pitched roof over 10° (1:6)

A small slope to flat roofs is necessary to encourage rainwater to flow towards rainwater gutters or outlets and to avoid the effect known as ponding. Where there is no slope or fall or a very shallow slope it is possible that rainwater may not run off the roof particularly at the centre of a roof where deflection under load has caused the roof to sink and cause rainwater to lie as a shallow pond. This water may in time cause deterioration of a membrane and consequent failure.

The minimum falls (slope) for flat roof coverings are 1:60 for sheet copper, aluminium and zinc and 1:80 for sheet lead and for mastic asphalt and built-up bitumen felt membranes.

To allow for deflection under load and for inaccuracies in construction it is recommended that the actual fall or slope should be twice the minimum and allowance made where slopes intersect so that the fall at the mitre of intersection is maintained. This would further increase the fall to say 1:28 which would require a considerable depth of firring for timber roof and depth of screeds or splay cut insulation boards to concrete roofs. Plainly the fall will depend on the size of the roof and arrangement of falls and outlets.

The traditional materials that are used as a weather covering for flat roofs, lead sheet, copper sheet and zinc sheet are laid as comparatively small sheets dressed over rolls at junctions of sheet down the slope of the roof and with small steps (drips) at junctions of sheet across the slope. The size of the sheet is limited to prevent excessive expansion and contraction that might otherwise cause the sheet to tear. The upstand rolls down the slope and the steps or drips across the slope provide a means of securing the sheets against wind uplift and as a weathering to shed water away from the laps between sheets. More recently aluminium sheet has also been used. Lead, copper and aluminium are not affected by normal weathering agents, do not progressively corrode and

have a useful life of very many years. Zinc is less durable than the three other non-ferrous metals used for roof covering.

Because of the cost of the material and the very considerable labour costs involved in covering a roof with one of these metals, the continuous membrane materials, asphalt and bitumen felt came into use early this century. Asphalt is a dense material that is spread while hot over the surface of a roof to form a continuous membrane which is impermeable to water. It is generally preferred as a roof covering to concrete flat roofs. Bitumen impregnated felt is laid in three layers bonded in hot bitumen to form a continuous membrane that is impermeable to water. Bitumen felt is comparatively lightweight and commonly preferred for timber flat roofs for that reason.

Both asphalt and bitumen felt oxidise and gradually become brittle and have a useful life of at most 20 years. Bitumen felt will become brittle in time and may tear due to expansion and contraction caused by temperature fluctuations. Of recent years fibre, glass fibre and asbestos fibre-based bitumen felts have lost favour due to premature failures and polyester-based felts are now more used for their greater resistance to tear and lower rates of oxidisation and embrittlement.

With the use of insulation under flat roof coverings the temperature fluctuations that the covering will suffer are considerable and the consequent strain on continuous coverings accelerates failures. Because of the rainfall common to northern Europe it would seem wise to avoid where possible, the use of flat roofs particularly for large areas of roof.

Sheet metal roof coverings

Sheet metal is used as a covering because it gives excellent protection against wind and rain, it is durable and is lighter in weight than asphalt, tiles or slates. Four metals in sheet form are used, namely lead, copper, zinc and aluminium.

Lead: A heavy metal which is comparatively soft and has poor resistance to tearing and crushing and has to be used in comparatively thick sheets as a roof covering. It is malleable and can easily be bent and beaten into quite complicated shapes without damage to the sheets. Lead is resistant to all weathering agents including mild acids in rainwater in industrially polluted atmospheres. On exposure to the atmosphere a film of basic carbonate of lead oxide forms on the surface of the sheets. These films adhere strongly to the lead and as they are non-absorbent they prevent further corrosion of the lead below them. The useful life of sheet lead as a roof covering is upwards of a hundred years.

Copper: A heavy metal which has good mechanical strength and is malleable. Because of its mechanical strength this metal can be used in quite thin sheets as a roof covering. The sheets of copper can readily be beaten and bent to quite complicated shapes. Like lead, on exposure to atmosphere a thin coat of copper oxide forms on the surface of copper sheets which is tenacious, non-absorbent and prevents further oxidisation of the copper below it. Copper is resistant to all normal weathering agents and its useful life as a roof covering is as long as that of lead.

Zinc: One of the lighter metals which has good mechanical strength but is not so malleable as lead or copper. In sheet form zinc can be bent and shaped but it tends to become brittle and break. On exposure to atmosphere a film of zinc oxide forms on the surface of the zinc sheets. This coating is not as dense or adherent as that which forms on lead or copper and gradually the zinc below corrodes to form zinc oxide. For this reason the useful life of sheet zinc as a roof covering is only twenty to forty years for the thickness of sheet usually employed. Zinc sheet is liable to damage in very heavily polluted industrial atmospheres and should not be used there. The cost of zinc sheet as a roof covering is less than that of lead or copper sheet and it is often used for that reason.

Aluminium: One of the lightest metals. It has moderate mechanical strength and is as malleable as copper. It is resistant to all normal weathering agents. On exposure to atmosphere a film of aluminium oxide forms which is dense and tenacious and prevents further corrosion. Aluminium as a roof covering has a useful life intermediate between zinc and lead.

Size of sheets: The size of metal sheet used for roof covering is determined by the size of sheets manufactured and the need to allow contraction and expansion of the sheet. If too large a sheet were used and fixed to a roof it might well tear due to contraction. The following is a table of the size of sheet commonly manufactured:

Lead: Rolls 2.4 wide and up to 12.0 long
Copper: Sheets size 1.2×600 and 1.8×900
Zinc: Sheets size 2.4×900
Aluminium: Sheets size 1.8×600, 1.8×900 and 1.8×1.2.

Jointing sheets: As has been mentioned, the sheets of metal have to be fixed to the roof and jointed so that they can expand and contract without tearing and three types of joint have been developed which successfully joint the sheets, keep out water and allow the sheets to contract and expand without tearing.

All sheet metals are laid to a fall or slope on roofs so that water runs off. For flat roofs the slope is as little as 1:80, and the joints across the fall or slope of the roof have to be designed to allow rainwater to run over them whereas the joints along the fall can project above the surface of the sheets of metal. The joints along or longitudinal to the fall are usually in the form of a roll. Rounded timber battens some 50 square are nailed to the roof and the edges of the sheets are either overlapped or covered at these timber rolls. The joints across or transverse to the fall of the roof are always formed as a small step called a drip. The purpose of the drip is to accelerate the flow of rainwater running down the shallow slope of the roof. The following notes and illustrations describe the use of the four metals as sheet coverings to flat roofs.

Sheet lead: The usual thickness of lead used for roof work is 1.8 mm, 2.24 mm, 2.5 mm, 3.15 mm or 3.55 mm. These thicknesses are described as Code No. 4, 5, 6, 7 and 8 respectively, the code number corresponding to Imperial weights of the sheet.

No sheet of lead should be larger than 1.6 m² so that the joints between the sheets are sufficiently closely spaced to allow the metal to contract without tearing away from its fixing. Another reason for limiting the size of sheet, which is peculiar to lead, is to prevent the sheet from creeping down the roof. The expression creep describes the tendency of the sheet to elongate. As the temperature of the metal rises the sheet expands but owing to its weight and poor mechanical strength it may not be able to fully contract as the temperature falls. The consequence is that the sheet gradually elongates over many years and becomes thinner and may in time let in water. It is not likely that this will happen on a flat roof with a sheet not larger than 1.6^2. The joints across the fall of the roof are made in the form of a 50 drip or step down and to reduce excessive increases in the thickness of the roof due to these drips they are spaced up to 2.3 apart and the rolls (joint longitudinal to fall) up to 800 apart. Figure 218 illustrates part of a lead covered flat roof showing the general layout of the sheets and a parapet wall around two sides of the roof.

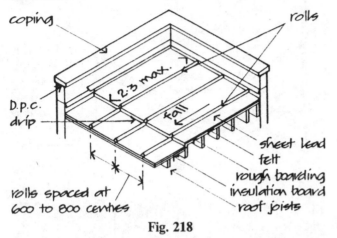

Fig. 218

Wood rolls. The edges of sheets longitudinal to the fall are lapped over a timber which is cut from lengths of timber 50 square to form a wood roll. Two edges of the batten are rounded so that the soft metal can be dressed over it without damage from sharp edges. Two sides of the batten are slightly splayed and the waist so formed allows the sheet to be clenched over the roll. Figure 219 is a view of a wood roll. An underlay of bitumen impregnated felt or stout waterproof building paper is first laid across the whole of the roof boarding and the wood rolls are then nailed to the roof at from 600 to 800 centres. The purpose of the underlay of felt or building paper is to provide a smooth surface on which the sheet lead can contract and expand.

The roof boarding on roofs to be covered with sheet lead may be fixed diagonally so that the joints between the boards are at 45 degrees to the fall. It is wasteful of timber to lay boards diagonally as the end of each board has to be cut off at 45 degrees and boards are, therefore, laid so that they run along the fall of the roof. The reason for laying the boards either diagonally or along the fall is so that if a board shrinks and warps it will not obstruct the flow of rain off the roof.

The edges of adjacent sheets are dressed over the wood roll in turn. In sheet metal work the word dressed is used to describe the shaping of the sheet.

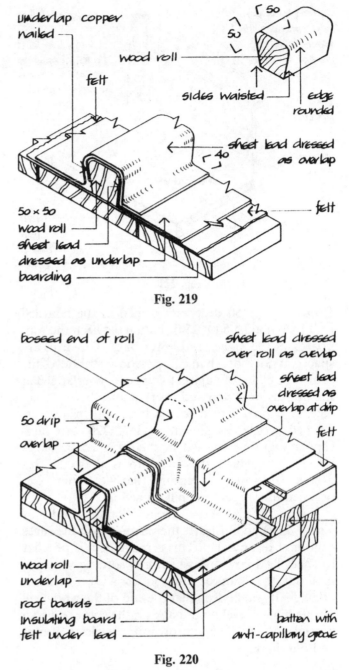

Fig. 219

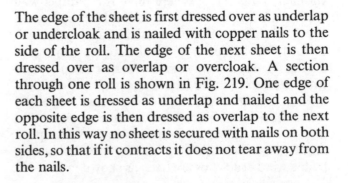

Fig. 220

The edge of the sheet is first dressed over as underlap or undercloak and is nailed with copper nails to the side of the roll. The edge of the next sheet is then dressed over as overlap or overcloak. A section through one roll is shown in Fig. 219. One edge of each sheet is dressed as underlap and nailed and the opposite edge is then dressed as overlap to the next roll. In this way no sheet is secured with nails on both sides, so that if it contracts it does not tear away from the nails.

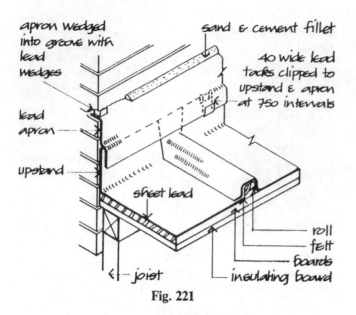

Fig. 221

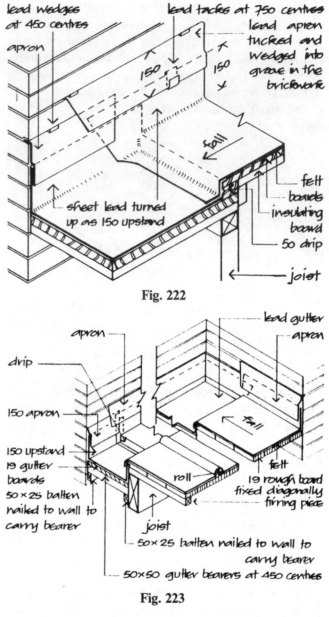

Fig. 222

Fig. 223

Drips. Drips 50 deep are formed in the boarded roof by nailing a 50 × 25 fir batten between the roof boards of the higher and lower bays. The drips are spaced at not more than 2.3 apart down the fall of the roof. The edges of adjacent sheets are overlapped at the drip as underlap and overlap and the underlap edge is copper nailed to the boarding in a cross-grained rebate, as shown in Fig. 220. An anti-capillary groove formed in the 50 × 25 batten is shown into which the underlap is dressed. This groove is formed to ensure that no water rises between the sheets by capillary action.

Figure 220 also shows the junction of wood rolls with a drip and illustrates the way in which the edges of four sheets overlap. This arrangement is peculiar to sheet lead covering which is a soft, very ductile material that can be dressed as shown without damage. It will be seen that the end of the wood roll on the higher level is cut back on the splay (called a bossed end) to facilitate dressing the lead over it without damage.

Upstand and apron. Where there is a parapet wall around the roof or where the roof is built up against a wall the sheets of lead are turned up against the wall about 150 as an upstand. The tops of these upstands are not fixed in any way so that the sheets can expand without restraint. To cover the gap between the upstand and the wall, strips of sheet lead are tucked into a horizontal brick joint, wedged in place and then dressed down over the upstand as an apron flashing. To prevent the apron from being blown up by the wind, lead clips are fixed as shown in Figs 221

and 222 which illustrate the junction of roll and drip with upstands.

Lead gutter. If the flat roof is surrounded on all sides by parapet walls it is necessary to collect the rainwater falling off at the lowest point of the roof. A shallow timber framed gutter is constructed and this gutter is lined with sheets of lead jointed at drips and with upstand and flashings similar to those on the roof itself. The gutter is constructed to slope or fall towards one or more rainwater outlets. The gutter is usually made 300 wide and is formed between one roof joist, spaced 300 from a wall, and the wall itself. Figure 223 shows the construction of the gutter.

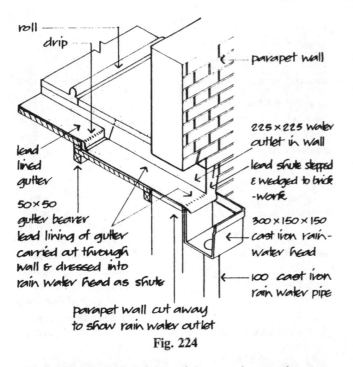

Fig. 224

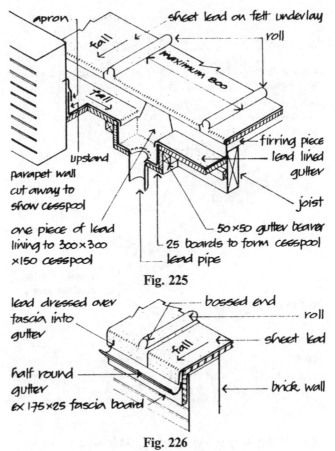

Fig. 225

Fig. 226

Rainwater outlet. If the rainwater pipe or pipes can be fixed on the outside of the building, the gutter discharges its rainwater through a lead shute in an opening in the parapet wall to a rainwater head and pipe. This construction is illustrated in Fig. 224.

If the rainwater pipe can only run down inside the building, it is usual to form a cesspool or catchpit at the end of the gutter. The cesspool acts as a reservoir so that during a heavy storm, when the rainwater pipe may not be able to carry the water away quickly enough, the cesspool prevents flooding of the roof. The construction of the cesspool is shown in Fig. 225. If there is no parapet to one side of the lead covered flat roof it will drain to a gutter fixed to a fascia board on the wall. This construction is shown in Fig. 226.

Copper sheet: The usual thickness of copper sheet for roofing is 0.6 mm. An oxide of copper forms on the surface of copper sheet and in the course of some few years the sheets become entirely covered with a light green compound of copper. This light green coating is described as a patina and is generally thought to give the copper sheets pleasing colour and texture. But in atmospheres heavily polluted with soot the patina is black instead of green. The patina, usually basic copper carbonate, is impervious to all normal weathering agents and protects the copper below it.

The standard sizes of sheet supplied for roofing are 1.2 × 200 and 1.8 × 900. The minimum fall for a copper covered roof is 38 in 3.0 and the fall is provided by means of firring pieces just as it is for lead covered roofs.

Rolls. The joints between sheets along (longitudinal to) the fall of the roof are formed over a wood roll fixed to the roof. Two sorts of joints are used (a) the batten roll and (b) the conical roll. Batten rolls are splay sided timber battens fixed to the roof at not more than 750 centres with brass screws, the heads of which are countersunk into the batten. The edges of the sheet are turned up each side of the batten and a separate strip of copper sheet is then welted to the roof sheets as a capping. A view of a batten roll is shown in Fig. 227.

It will be seen from Fig. 227 that the sheets are secured to the roof by means of copper cleats. These 50 wide strips of sheet are fixed under the rolls at not more than 450 apart and are folded in with the sheets and capping.

Instead of covering the wood roll with a separate capping the edges of the sheets can be folded together in the form of a double welt over a conical section roll. The rolls are fixed to the roof at not more than

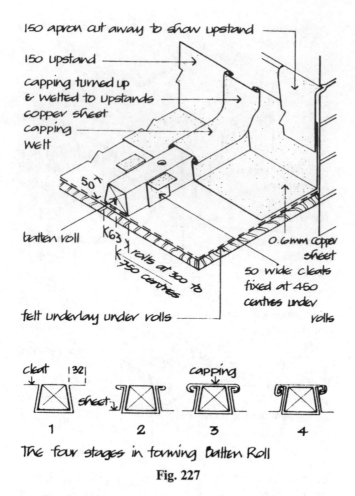

The four stages in forming Batten Roll

Fig. 227

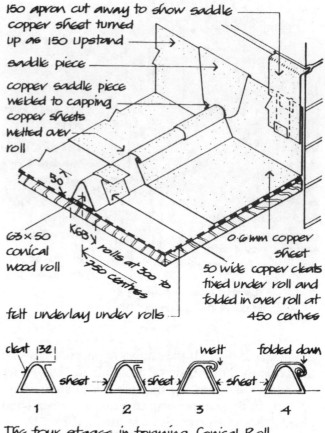

The four stages in forming Conical Roll

Fig. 228

750 centres. The arrangement of the copper sheets is shown in Fig. 228., and it will be seen that less sheet is required to form the conical roll than the batten roll joint.

Double lock cross welts. The drips formed in a roof covered with copper sheet are spaced up to 3.0 apart and their purpose is to accelerate the flow of water running down the fall of the roof. But as copper sheets are either 1.2 or 1.8 long some joint transverse to the fall has to be used between drips. This joint takes the form of a 'double lock welt' and because the joint is across the fall it is called a double lock cross welt. This joint is illustrated in Fig. 229.

It will be seen from Fig. 229 that the double lock cross welt is folded up with the sheets at rolls. To avoid too great a thickness of sheets the double lock cross welts are staggered as shown in Fig. 230 which illustrates a view of part of a copper covered flat roof.

Drips. These are formed at not more than 3.0 apart and a 63 or 70 step down is formed in the timber roof.

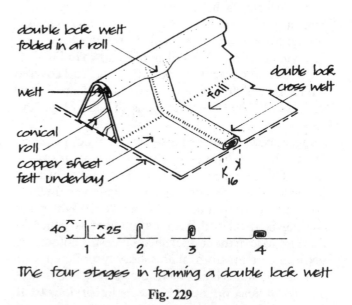

The four stages in forming a double lock welt

Fig. 229

The edges of the sheets are welted as shown in Fig. 231 which also illustrates the junction of the conical roll and a drip.

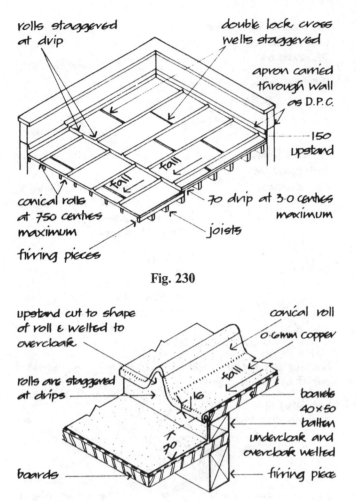

rolls staggered at drip

double lock cross welts staggered

apron carried through wall as D.P.C.

150 upstand

fall

fall

fall

conical rolls at 750 centres maximum

firring pieces

70 drip at 3·0 centres maximum

joists

Fig. 230

upstand cut to shape of roll & welted to overcloak

conical roll

0·6mm copper

fall

rolls are staggered at drips

boards 40 × 50 batten

16

undercloak and overcloak welted

70

boards

firring piece

Fig. 231

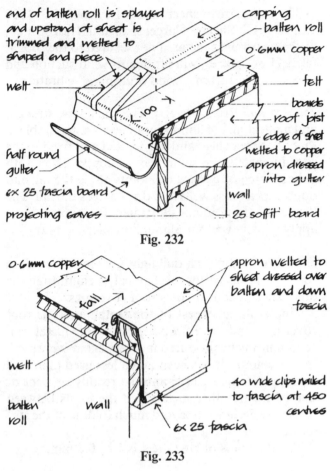

end of batten roll is splayed and upstand of sheet is trimmed and welted to shaped end piece

capping

batten roll

0·6mm copper

welt

100

felt

boards

roof joist

edge of sheet welted to copper

apron dressed into gutter

half round gutter

6 × 25 fascia board

projecting eaves

wall

25 soffit board

Fig. 232

0·6mm copper

apron welted to sheet dressed over batten and down fascia

fall

welt

batten roll

wall

40 wide clips nailed to fascia at 450 centres

6 × 25 fascia

Fig. 233

Upstand and apron. Where the roof is surrounded by a parapet wall or it adjoins a higher wall the copper sheet is turned up as an upstand average 150 high and this upstand is covered with an apron flashing as illustrated in the previous drawings of batten roll and conical roll.

Box gutter. Where the roof is surrounded by a parapet wall the rainwater is collected in a box gutter formed in the timber roof in exactly the same way as that formed for a lead covered roof. The gutter is usually 300 wide and drips at not more than 3.0 intervals are formed. The rainwater discharges either through the parapet wall to a rainwater head or to a cesspool as for lead covered flat roofs. The gutter is constructed as illustrated for lead lined gutters and the copper lining to it is jointed with drips.

Upstands and flashing are formed just as they are on the main part of the copper flat roof.

Eaves gutter. As copper sheet is thought to be an attractive roof covering the roof is not usually hidden behind a parapet wall and it is constructed to slope towards and discharge into an eaves gutter. Figure 232 is an illustration of the construction of the roof and the jointing of the sheets at such a gutter.

The arrangement of the batten roll end at eaves is identical to the batten roll end at a drip.

Verge. Where there is no parapet wall around the roof the edges or verges of the roof down to the fall are usually finished with a roll and a flashing fixed to a fascia board with copper clips as shown in Fig. 233.

Aluminium sheet: Aluminium sheet has a bright silver grey colour which in time gradually darkens. Pure aluminium containing at least 99% aluminium is used for sheet metal roof coverings. The sheets have moderate mechanical strength and can be readily bent and beaten into quite complicated shapes without damage. The sheets are usually 0.7 mm thick.

The usual size of sheet used is 1.8 × 600, 1.8 × 900 and 1.8 × 1.2. Sheets of larger size should not be used, as the lightweight metal, if not secured to the roof at fairly close centres, will lift due to wind suction and may cause disturbing 'wind drumming' vibration.

Jointing and fixing sheets. The thickness, strength and malleability of aluminium sheet is comparable to that of copper sheet and it is jointed and fixed to the roof in exactly the same way as copper sheet with either batten or conical roll joints along the fall and double lock cross welts and drips across the fall. The details of these joints illustrated for copper sheet apply equally well for aluminium sheet.

Zinc sheet: Zinc is a dull light grey metal which in sheet form has better mechanical strength than the other sheet metals used as roof coverings.

Zinc is the cheapest of the metals used as roof coverings and in Europe, where the metal can economically be produced it is very extensively used. In this country it has been much less used than lead or copper. Zinc sheet cannot so readily be bent or shaped as the other metals and the joints between sheets are designed to avoid much folding of the stiff metal.

The thickness of zinc sheet is 1 or 0.8 mm.

Jointing and fixing sheets. The standard sheet is 2.4 by 900. The joints along the fall of the roof are formed over a wood batten. The sheets are bent up on either side of the batten which is covered with a zinc capping as shown in Fig. 234.

The sheets are secured by means of sheet zinc clips 40 wide which are nailed under the battens at 750 centres and are turned up and clipped over the edge of the sheets as shown in Fig. 234. The zinc capping is secured by means of a holding down clip which is folded out of a strip of a zinc sheet. The zinc capping is made in 1.35 lengths. A length of capping is placed on the batten and the holding down clip is then nailed to the batten over the end of the capping as shown in Fig. 235.

The end of the next length of capping is inserted into the fold in the holding down clip and then placed on the batten and in turn secured with a holding down clip.

Using a standard 900 wide sheet the battens are fixed at 850 centre to centre. Using the standard 2.4 long sheet, drips are formed at 2.3 intervals down the fall of the roof and the following is an illustration of

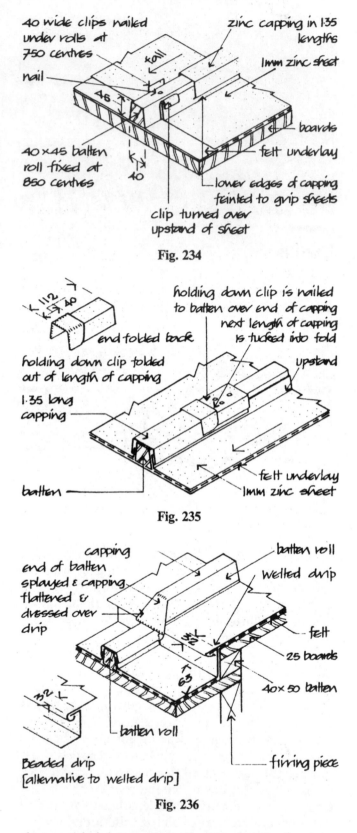

Fig. 234

Fig. 235

Fig. 236

the two drip joints commonly used, and details of the junction of the rolls and drip, Fig. 236.

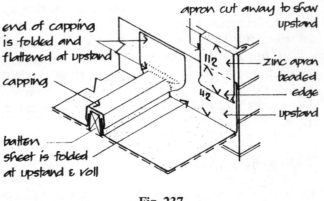

end of capping is folded and flattened at upstand

capping

batten

sheet is folded at upstand & roll

apron cut away to show upstand

112

42

zinc apron beaded edge

upstand

upstand

Fig. 237

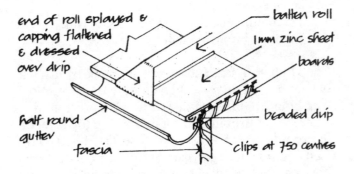

end of roll splayed & capping flattened & dressed over drip

half round gutter

fascia

batten roll

1mm zinc sheet

boards

beaded drip

clips at 750 centres

Fig. 238

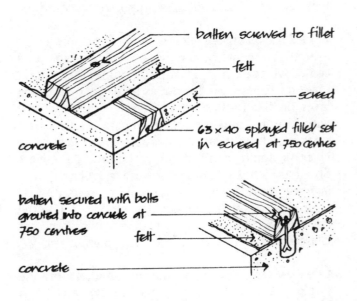

batten screwed to fillet

felt

screed

63 × 40 splayed fillet set in screed at 750 centres

concrete

batten secured with bolts grouted into concrete at 750 centres

felt

concrete

Fig. 239

Where there is a parapet wall around the roof the sheets are turned up against it 150 as an upstand and the upstand is covered with an apron flashing tucked into a joint in the brickwork as shown in Fig. 237.

If the roof is surrounded by a parapet wall it drains to a zinc lined box gutter similar to that used with other sheet metal roof coverings, only lined with zinc sheet. If there is no parapet wall the sheets drain directly on to an eaves gutter as shown in Fig. 238.

Sheet metal covering to concrete roofs: Because they are cheaper in first cost and easy to lay, bitumen felt and asphalt have principally been used as a covering to concrete flat roofs. But felt and asphalt only have a useful life of some twenty years and one of the sheet metals is sometimes used instead. The sheet metal is jointed and fixed to a concrete roof in the same way as on a timber roof. The wood rolls are secured to the concrete by screwing them to splayed timber battens set into the screed on the concrete or by securing them with bolts set in sand and cement in holes punched in the screed as shown in Fig. 239.

Drips should be formed in the surfaces of the roof just as they are in the surface of a timber roof and the details of jointing and dressing of the sheets is the same as those illustrated for timber roofs.

Roofing felt: It is essential that sheet metal be laid on a continuous layer of roofing felt laid on the surface of the concrete roof. The felt enables the metal to contract and expand freely and prevents it tearing on any sharp projections in the surface of the concrete roof.

Waterproof membranes

Mastic asphalt

Asphalt (sometimes spelt asphalte) is defined as a mixture of naturally occurring material which is soft, has a low melting point and is an effective barrier to the penetration of water.

Natural rock asphalt is mined from beds of limestone which were saturated, or impregnated, with asphaltic bitumen thousands of years ago. The rock is chocolate brown in colour and is mined in several districts around the Alps and Europe. The rock is hard and because of the bitumen with which it is impregnated it does not as readily absorb water as ordinary limestone.

Natural lake asphalt is dredged principally from

the bed of a dried up lake in Trinidad. It contains a high percentage of bitumen with some water and finely divided solid material.

Asphalt is manufactured either by crushing natural rock asphalt and mixing it with natural lake asphalt, or by crushing natural limestone and mixing it with bitumen whilst the two materials are sufficiently hot to run together. These materials are defined as *Mastic Asphalt for Roofing (Natural Rock Asphalt Aggregate)* BS 6577 and *Mastic Asphalt for Roofing (Limestone Aggregate)* BS 988. The heated asphalt mixture is run into moulds in which it solidifies as it cools.

The solid blocks of asphalt are heated on the building site and the hot plastic material is spread over the surface of the roof in two layers breaking joint to a finished thickness of 20. As it cools it hardens and forms a continuous, hard, waterproof surface. If the roof has parapet walls around it, or adjoins the wall of a higher building, an asphalt skirting or upstand, 150 high, is formed and this skirting is turned into horizontal brick joints purposely cut about 25 deep to take the turn-in of the asphalt skirting. It is essential that asphalt be laid on an isolating membrane underlay of black haired felt so that slight movements in the structure are not reflected in the asphalt membrane. A properly laid asphalt covering to a roof will not absorb water at all and the finished surface of the asphalt can be absolutely flat as any rainwater that lies on it will eventually evaporate. But it is usual to construct flat roofs so that the asphalt is laid to a slight fall, of at least 1:80, so the rainwater drains away to a rainwater outlet or gutter. Asphalt is a comparatively cheap roof covering and if the asphalt is of good quality and is properly laid it will have a useful life of some 20 years or more. The asphalt should be renewed about every 20 years if the roof is to be guaranteed watertight.

If there is no parapet wall around the roof it is usually designed to overhang the external walls, to give them some protection, and the asphalt drains to a gutter. Figure 240 illustrates simple flat roofs constructed with timber or concrete with and without parapet walls, and covered with asphalt.

Cement screeds and particularly lightweight aggregate screeds on concrete roofs take time to thoroughly dry out and may absorb rainwater so that it is highly likely that some water will be trapped in the screed once asphalt or bitumen felt covering has been applied. The heat of the sun will then cause

this water to vaporise and the vapour pressure may cause the covering to blister, crack and let in water. To relieve this water vapour pressure, it is practice to use a venting layer of felt or partial bonding on wet screeded roofs. The venting layer allows water vapour to be released through vapour pressure releases at abutments and verges or through vents in the roof.

Built-up bitumen felt roofing

There are five classes of bitumen felt included in BS 747 as:

Class 1 bitumen felts (fibre base)
Class 2 bitumen felts (asbestos base)
Class 3 bitumen felts (glass fibre base)
Class 4 sheathing felts and hair felts
Class 5 bitumen felts (polyester base) with
    oxidised bitumen coating

Class 1 bitumen felts (fibre base) are made from a felt base of animal or vegetable fibres that are saturated with bitumen. Fine granule surfaced felts are used as a lower layer in built-up roofing and as a top layer on flat roofs that are subsequently surfaced with bitumen and mineral aggregate finish. Mineral surfaced fibre base felt is finished on one side with mineral granules for appearance and protection for use as a top layer on sloping roofs. Reinforced fibre based felts have a layer of jute hessian embedded in the coating on one side and are used under slates and tiles where the felt is not fully supported by boarding.

Class 2 and 3 bitumen felts are made respectively with asbestos fibre and bonded glass fibre saturated and coated with bitumen. The fine granule coated felts in these two classes are for use as underlay or as a top layer on flat roofs that are subsequently surfaced with bitumen and mineral aggregate. The mineral surfaced felts in these two classes are for use as a top layer on sloping roofs. A third type of Class 3, glass fibre base felt is perforated for use as a venting first layer on roofing when partail bonding is used.

Because of the health hazards associated with the use of asbestos fibre in manufacturing it is likely that Class 2 felts will be withdrawn before long.

Class 4 sheathing and hair felts are made from long staple fibres, loosely felted and impregnated with bitumen (black felt) or brown wood tars or wood

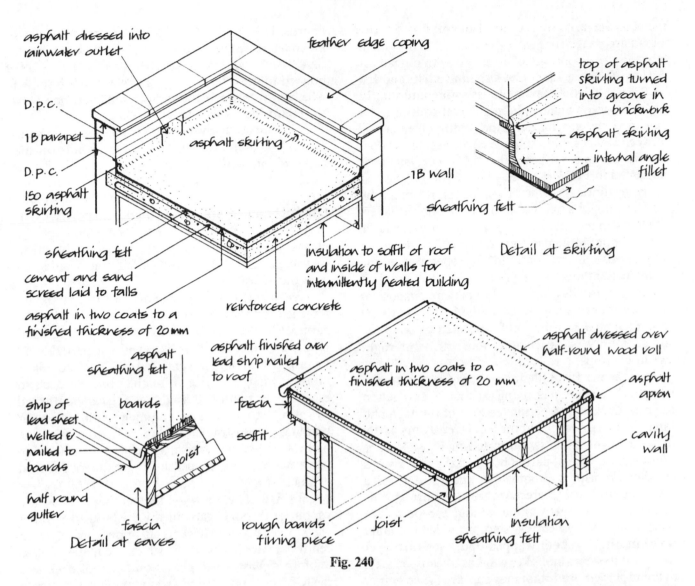

asphalt dressed into rainwater outlet

feather edge coping

top of asphalt skirting turned into groove in brickwork

D.p.c.

1B parapet

D.p.c.

asphalt skirting

150 asphalt skirting

asphalt skirting

internal angle fillet

1B wall

sheathing felt

sheathing felt

cement and sand screed laid to falls

insulation to soffit of roof and inside of walls for intermittently heated building

Detail at skirting

asphalt in two coats to a finished thickness of 20 mm

reinforced concrete

asphalt finished over lead strip nailed to roof

asphalt dressed over half-round wood roll

asphalt sheathing felt

asphalt in two coats to a finished thickness of 20 mm

asphalt apron

boards

fascia

strip of lead sheet welted & nailed to boards

soffit

cavity wall

joist

half round gutter

fascia

rough boards tilting piece

joist

insulation

sheathing felt

Detail at eaves

Fig. 240

pitches (brown felts). These felts are used as underlays for mastic asphalt roofing and flooring to isolate the asphalt from the substructure.

Class 5 bitumen felts (polyester base) with oxidised bitumen coating are made with a non-woven fabric base of polyester staple fibre or polyester filament, formed by needling or spin bonding, that is impregnated with bitumen. The fine granule surfaced felt is used as a base, intermediate or top layer which is to be subsequently covered with bitumen and mineral aggregate finish and the mineral surfaced felt as a top layer on flat roofs without additional surface treatment.

The purpose of the top dressing of bitumen and mineral aggregate to fine granule surfaced felts on flat roofs is to act as a reflective and protective layer.

The mineral aggregate is of a light coloured mineral graded in size from 16 to 32 and spread to a thickness of at least 12.5 mm.

Class 5 felts were introduced a few years ago. They have higher tensile strength than other classes of felt and are better able to withstand the strains that a flat roof covering will suffer, without rupture. It is common practice today to use Class 1, 2 and 3 felts for low cost roofing and small-scale domestic work, such as extensions and to use Class 5 and other polyester base felts for all other work. Because of the higher tensile strength of polyester base felts they are sometimes referred to as High Performance or HT membranes.

Laying bitumen felt roofing: Bitumen felt is applied to flat and sloped roofs in three layers the first,

175

intermediate and top layers. The rolls of felt are spread across the roof with a side lap of 75 minimum between the long edges of rolls and with a head, or end lap of at least 75 for Class 1, 2 and 3 felts and 150 for Class 5 felts. To avoid an excessive build up of thickness of felt at laps, the side lap of rolls of felt is staggered by one-third of the width of each roll between layers so that the side lap of each layer does not lie below or above that of other layers as illustrated in Fig. 241.

The traditional method of fixing felt roofing is by fully bonding the felt to the roof deck and the layers of felt to one another by the 'pour and roll' technique in which hot bitumen is poured on the roof deck and the rolls of felt are continuously rolled out as the bitumen is poured. The pour and roll method of bonding is used for all three layers. The purpose of the bitumen is initially to bond the first layer to the roof deck against wind uplift and then to bond the succeeding layers to each other and form a watertight seal at overlaps of rolls of felt.

A departure from the full bonding method is the partial bond method of fixing the first layer of felt to the roof deck with a perimeter strip of bitumen and intermediate strips so that the first layer is only partly bonded to the roof, as illustrated in Fig. 241. The principal purpose of partial bonding the first layer is to allow moisture vapour that may rise from a concrete roof and screed to move through the unbonded spaces to ventilators and escape to air. Where this sort of venting is not provided the build up of moisture vapour trapped under the first layer may turn to steam and so expand as to cause blisters in the roof membrane that may rupture the covering.

The bond of the first layer of felt to the roof deck depends on the surface to which it is to be applied. On a surface of timber boarding nailed to firring or through insulation, the first layer is fixed to the boards with clout nails at 150 staggered centres over the area of each sheet and at 50 staggered centres on the centre line of all overlaps of sheets. The purpose of nailing to boards is that the boards may move to some extent by drying out, may lose shape in drying and so not provide a sound level base for a bitumen bond. The layers above are then fully bonded with bitumen. Fig. 241 is an illustration of felt applied to a timber boarded flat roof.

On a roof deck covered with sheets of plywood or particle board the joints between the boards are first covered with strips of self adhesive tape and then a perforated first layer of felt is laid loose over the boards. This first layer of felt is usually of fibre glass-based felt which is perforated with holes as a venting first layer. The intermediate layer is then fully bonded to the loose venting layer with hot bitumen which runs through the perforations and bonds to the boards below as partial bonding.

On a roof deck covered with insulation boards the method of bonding the first layer depends on the composition of the insulation. The majority of insulation boards either have a surface to which hot bitumen can be applied or are coated to assist the bond of bitumen. On polyurethane and polyisocyanurate boards, provision should be made for the escape of gases generated by the use of hot bitumen, by using a venting first layer of felt that is laid loose over the boards.

On concrete flat roofs that are cast in-situ and on screeded finishes to pre-cast concrete roofs the surface should first be sealed with a sealing compound and then the felt is fixed with the partial bond method by laying a venting first layer to allow moisture that may rise from the concrete deck to escape to ventilators. Partial bonding can be effected by using strips of hot bitumen with perimeter bonding as illustrated in Fig. 241, or one of the perforated venting layers may be used with 500 wide perimeter bonding for attachment to the roof against wind uplift.

On small concrete roofs the venting of moisture vapour is through gaps in the upstand to parapet walls. On large roofs vents at approximately 6 m centres are used. The vents, which are made of glass reinforced plastic, are placed on the roof deck and the felt is cut and dressed around them as illustrated in Fig. 242.

For most permanent work the suppliers of roofing felts recommend the use of Class 5 polyester-based felts for the three layers of built-up bitumen felt roofing, or one of their polyester based HT felts and Class 3 glass fibre-based felt for a venting first layer for partial bonding. There is some small saving in the use of glass fibre-based felt for the intermediate layer below a Class 5 or HT polyester-based felt.

On flat roofs the felt should be finished with a top dressing of white mineral chippings spread over a bitumen bonding compound. The chippings should be spread to a thickness of at least 12.5 mm. The mineral chippings serve to reflect the suns rays, absorb some of the heat generated by the energy from sunlight, give protection against spread of flame and serve as a finish for appearances sake.

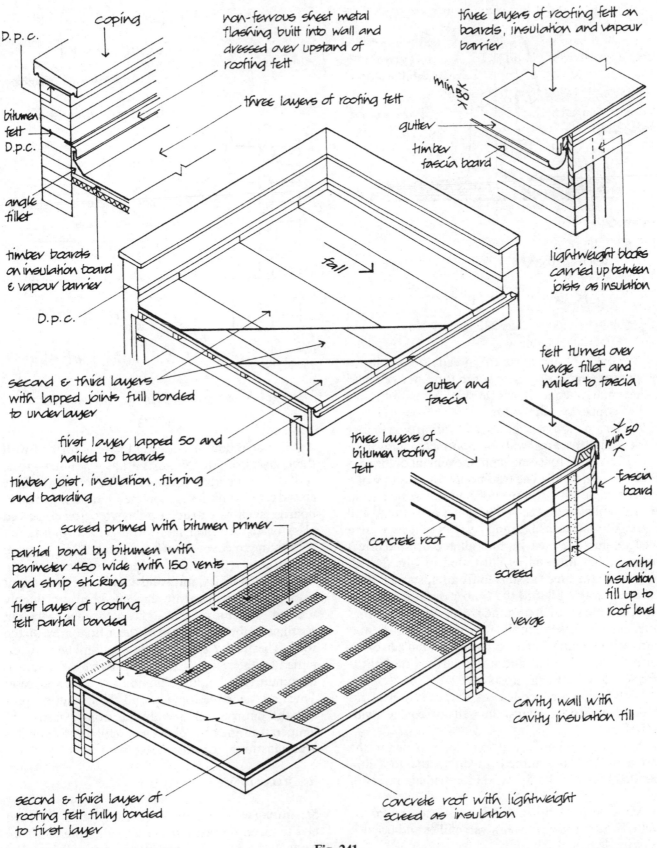

coping

D.p.c.

non-ferrous sheet metal flashing built into wall and dressed over upstand of roofing felt

three layers of roofing felt on boards, insulation and vapour barrier

bitumen felt

D.p.c.

three layers of roofing felt

gutter

min 50

timber fascia board

angle fillet

fall

timber boards on insulation board & vapour barrier

D.p.c.

lightweight blocks carried up between joists as insulation

second & third layers with lapped joints full bonded to underlayer

gutter and fascia

felt turned over verge fillet and nailed to fascia

first layer lapped 50 and nailed to boards

three layers of bitumen roofing felt

min 50

timber joist, insulation, firring and boarding

fascia board

screed primed with bitumen primer

concrete roof

partial bond by bitumen with perimeter 450 wide with 150 vents and strip sticking

screed

cavity insulation fill up to roof level

first layer of roofing felt partial bonded

verge

cavity wall with cavity insulation fill

second & third layer of roofing felt fully bonded to first layer

concrete roof with lightweight screed as insulation

Fig. 241

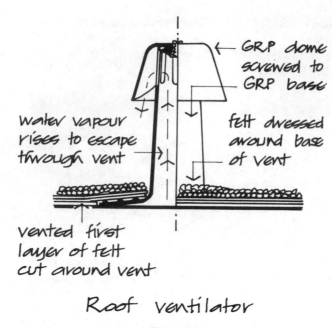

Roof ventilator

Fig. 242

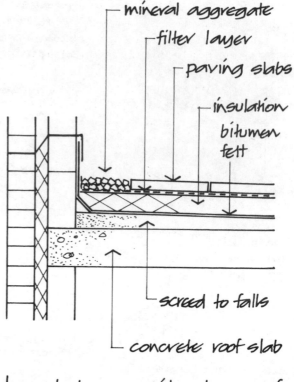

Inverted or upside down roof

Fig. 243

As it is not practical to spread loose mineral chippings over a sloped roof the top layer of felt is either mineral coated or is finished with a reflective foil of copper or aluminium.

At the junction of built-up felt roofing on a timber deck and a parapet wall or abutment, allowance should be made for some slight movement of the roof relative to the wall. The traditional finish is to nail a timber angle fillet to the roof and dress the felt up over the fillet and against the wall as an upstand 150 high. A non-ferrous flashing, usually of lead, is then wedged into a groove in the wall and dressed down over the felt upstand as illustrated in Fig. 241. A more sophisticated finish is to fix an upstand timber curb and angle fillet to the timber deck, free of the wall, to which the felt upstand is dressed. The top and intermediate layers are then secured to the upstand curb with a batten of wood which is nailed to the curb through the felt and a non-ferrous flashing is dressed down over the upstand.

With felt on a concrete deck the felt is dressed up over an angle fillet against the wall and covered with a flashing.

At eaves to a gutter, a strip of felt is nailed to the fascia, welted and turned up on to the roof and bonded between the first and intermediate layers of felt.

At verges, a strip of felt is nailed to the fascia, welted and turned up over a verge fillet and bonded between layers of felt.

An upside down or inverted roof is formed with three layers of felt fully or partially bonded to the roof deck depending on its construction. One of the closed cell organic insulation materials in the form of boards, such as extruded polystyrene or coalesced glass beads are then laid loose on top of the felt membrane to cover the whole of the roof. An open weave fabric is then laid over the top of the insulation to separate the insulation and aggregate finish. Light coloured stone aggregate that has been thoroughly washed to remove dust, clay or other adherent coatings is then spread over the whole area of the roof. The thickness of the stone topping is determined by the weight required to resist the assumed maximum wind uplift that may act on the roof. Together the stone topping and the insualtion protect the bitumen felt membrane from extremes of temperature and so prolong its useful life. Figure 243 is an illustration of an inverted roof.

Resistance to fire

Separating walls, party walls, between semi-detached and terraced houses should resist the spread of fire from one house to the next, by being raised above the level of the roof covering at least 375 or by being built up to the underside of the roof covering on timber roofs and the underside of concrete roofs. These walls must serve as a continuous fire break up to the underside of the concrete roof or underside of decking to covering on timber roofs through solid mortar bedding between the top of the wall and the roof or by sealing with fire stopping material. Where bitumen felt roofing is laid over a separating wall the roof deck should have AA, AB or AC designation for external fire exposure for 1.5 each side of the separating wall faces. This is a precaution to delay the spread of fire through the roof of one house to the next.

Resistance to the passage of heat

A flat roof provides poor insulation against transfer of heat. Most of the materials used in the construction of roofs are poor insulators and some material has to be built into or on to flat roofs to improve insulation to a maximum U value of 0.25 to meet the 1989 revision of the requirements of The Building Regulations 1985.

The most practical position for the layer of insulation required for a flat roof is on top of the roof deck either below or over the weathering membrane or cover. This warm roof or warm construction avoids the need for ventilation to timber flat roofs and utilises the heat store capacity of concrete roofs where continuous low output heating is used inside buildings. Any one of the insulants may be used in the form of boards that are laid over the top surface of the roof deck. The insulating boards may be formed or cut to provide the necessary fall or slope for the covering, providing the insulating thickness at the lowest point is sufficient to provide minimum insulation values. Details of materials used for insulation of warm flat roofs are set out in table 18.

To minimise thermal, cold bridges at junctions of roof and walls against loss of heat and possible condensation, it is important to unite or continue wall insulation up to that of the roof. With insulation on top of a roof deck the cavity wall insulation should be continued up in the parapet wall to the level of the d.p.c. and a low density block be used for the inner leaf to minimise a cold bridge through the parapet and roof deck as illustrated in Fig. 244. With eaves finished flush with the outside face of a wall the roof top insulation should be united with cavity insulation as illustrated in Fig. 245, and inside

Table 18. Insulation Materials

Flat roof warm roof	Thickness	U valve W/m²K
Rockwool slab 1200 × 60	30, 40, 50, 60, 70, 80, 90, 100	0.036
slab cut to falls 1200 × 890	30, 40, 50, 60, 70, 80, 90, 100	0.036
Cellular glass board 1200 × 600	30, 35, 40, 50, 60, 70, 80	0.042
EPS board uniform in thickness or tapered	from 20 up in 5 increments	0.034
XPS board 1250 × 600	50, 75, 90, 100, 120	0.025
PIR board faced with glass fibre tissue 1200 × 750	35, 40, 50	0.022
PUR board bitumen impregnated glass fibre tissue faced 1200 × 75	26, 32, 35, 40, 50, 75	0.022
PUR board glass tissue fibre faced 2400 × 1200	25, 30, 35, 50, 55	0.022

EPS expanded polystyrene
XPS extruded polystyrene
PIR rigid polyisocyanurate
PUR rigid polyurethane

insulation to a wall should be continuous with that of a roof as illustrated in Fig. 245. Where the roof deck projects beyond the face of the external wall as an eaves overhang, the insulation should be continued from roof level down the vertical face and across the underside of the roof overhang to unite with wall insulation as illustrated in Fig. 246. This can be effected by fixing insulation board to the roof timbers behind the fascia and soffit board of timber roofs and by fixing insulation board to concrete with adhesive, by non-ferrous fixing or by using the insulation as permanent formwork. The insulation is then covered with fascia and soffit boards or it is rendered.

To prevent moisture vapour from warm, moist inside air penetrating the roof deck and insulation and condensing on the cold side of insulation it is necessary to form a vapour check below insulants that are permeable to moisture vapour.

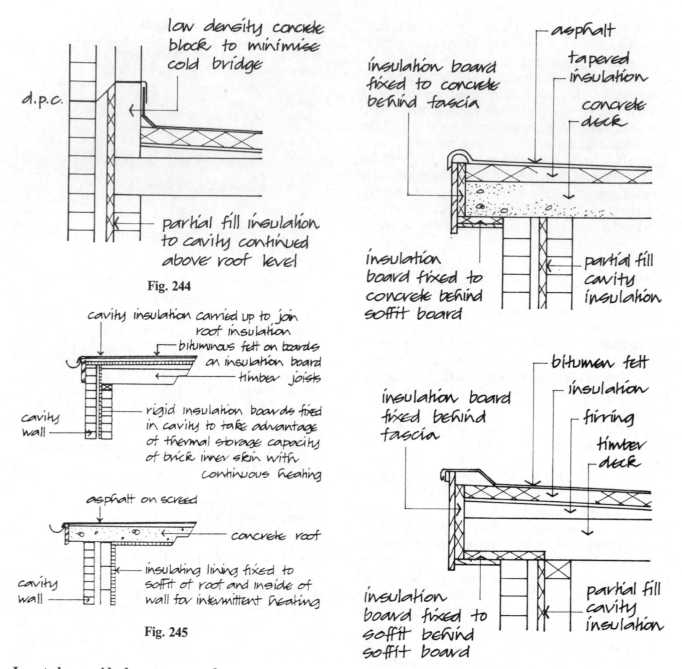

low density concrete block to minimise cold bridge

d.p.c.

partial fill insulation to cavity continued above roof level

Fig. 244

cavity insulation carried up to join roof insulation

bituminous felt on boards on insulation board

timber joists

cavity wall

rigid insulation boards fixed in cavity to take advantage of thermal storage capacity of brick inner skin with continuous heating

asphalt on screed

concrete roof

cavity wall

insulating lining fixed to soffit of roof and inside of wall for intermittent heating

Fig. 245

asphalt

tapered insulation

concrete deck

insulation board fixed to concrete behind fascia

insulation board fixed to concrete behind soffit board

partial fill cavity insulation

bitumen felt

insulation

firring

timber deck

insulation board fixed behind fascia

insulation board fixed to soffit behind soffit board

partial fill cavity insulation

Edge insulation to projecting eaves of flat roofs

Fig. 246

Inverted or upside down warm roof

To provide a layer of insulation to the roof and also to protect roof membranes from the very considerable temperature fluctuations that a roof surface will suffer, the inverted or upside down roof is used. This is a form of warm roof with the insulation laid on top of the weather membrane as illustrated in Fig. 243. One of the closed cell organic materials or coalesced glass beads are used as an insulant because the insulation is exposed to rain which would otherwise soak into the material and greatly reduce its effect as

an insulant. To protect the top level insulation against wind uplift and to provide some reflective protection, the insulation is covered with mineral aggregate of sufficient weight to hold it down. Rain falling on the roof drains through the stone covering down to the insulation and across or through insulation to rainwater outlets in the weather membrane. The considerable surface area of the stone particles of the covering tend to collect and hold water during spells of heavy rain which delays run off towards outlets and so minimises the possibility of flooding. The one small disadvantage of this form of roof is that it is impossible to examine the weather membrane for faults and that the stone covering has to be removed for repair or renewal.

Where the insulating material is fixed below a concrete flat roof, the wall insulation should be continued up to at least the underside of the deck to minimise cold bridges, and insulants that are pervious to moisture vapour should be covered with a vapour check on the warm side.

Where insulation is fixed between the joists of a timber flat roof or the insulation is fixed under a timber flat roof it is a requirement of The Building Regulations 1985 that there should be reasonable provision of ventilation in the roof void to prevent excessive condensation. In the current regulations it is proposed that cross ventilation be provided by ventilation inlets equivalent to a 25 continuous gap on opposite sides of timber roofs as illustrated in Fig. 247. The ventilation can be provided in the soffit of projecting eaves in the form of grilles that are covered with fine mesh to exclude bees and wasps. Where the insulation is fixed between joists there should be a clear space over the insulation of at least 50 to allow for air circulation. To reduce the penetration of moisture vapour to the void in these roofs there should be a vapour check on the warm side of insulants that are permeable to moisture vapour.

Details of materials used for cold, timber framed roofs are set out in table 19.

The method of determining the required thickness of insulation for flat roofs is the same as that for pitched roofs as set out in the 1989 revision of Approved Document L to The Building Regulations 1985. Insulation thickness may be determined by the conductivity of the insulant as set out in table 16 or by a calculation similar to that for pitched roofs by taking account of the thermal resistance of the covering and its surface, air space and plasterboard lining and its inside surface.

PARAPET WALLS

External walls of buildings are raised above the level of the roof as parapet walls for the sake of the

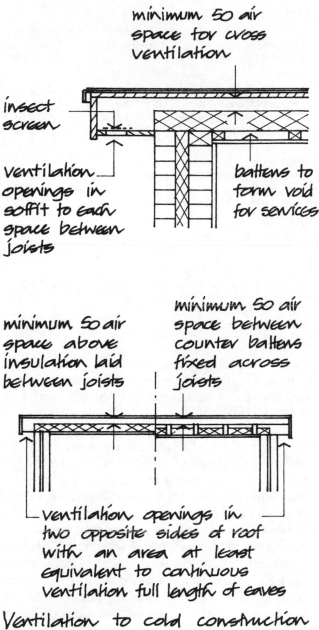

Ventilation to cold construction timber flat roof

Fig. 247

appearance of the building as a whole. Parapet walls are exposed on all faces to driving rain, wind and frost and are much more liable to damage than external walls below eaves level.

Because parapet walls are freestanding their height is limited in relation to their thickness for the sake of stability. Approved Document A, giving practical

Table 19. Insulating Materials

Flat roof cold roof	Thickness	*U* valve W/m²K
Glass fibre rolls between rafters semi-rigid batt friction fit between rafters 570 or 370	80, 100, 150 80, 90, 100, 120, 140, 160, 180, 200	0.037 0.036
Rockwool rolls between rafters semi-rigid batt friction fit between rafters	80, 100, 150 80, 90, 100, 120, 140, 180, 200	0.04 0.04
XPS board fixed to soffit	50, 75, 90, 100, 120	0.025
PIR board fixed to soffit	35, 40, 50	0.022

XPS extruded polystyrene
PIR rigid polyisocyanurate

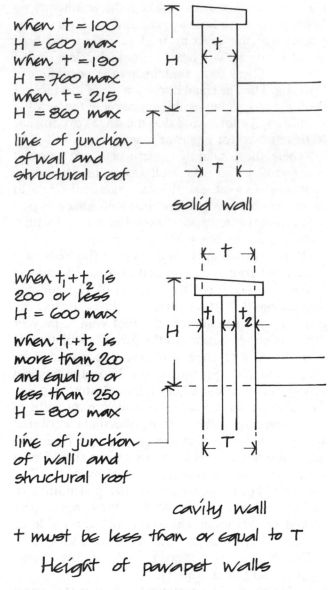

when t = 100
H = 600 max
when t = 190
H = 760 max
when t = 215
H = 860 max
line of junction of wall and structural roof

solid wall

when $t_1 + t_2$ is 200 or less
H = 600 max
when $t_1 + t_2$ is more than 200 and equal to or less than 250
H = 800 max
line of junction of wall and structural roof

cavity wall
t must be less than or equal to T
Height of parapet walls

Fig. 248

guidance to the requirements of The Building Regulations 1985 for houses and other small buildings, sets limits to the thickness and height of parapet walls as set out in Fig. 248.

The top surface of a parapet wall is exposed directly to rain and to prevent water saturating the wall it is essential that it should be covered or capped with some non-absorbent material. Natural stone was commonly used for this purpose as it is at once protective and decorative. The stones are described as coping stones and are cut so that they have a sloping top surface when laid and it is said that the stones are weathered. The stones usually project some 50 or more each side of the parapet wall so that rainwater running from them drips clear of the face of the wall. The three most usual sections employed for coping stones are shown in Fig. 249.

From the above sections it will be seen that a groove is cut on the underside of each stone where it overhangs the parapet. The purpose of these grooves is to prevent rainwater running in along the underside of the stones to the face of the wall. When the water reaches the groove it is unable to travel any further and so drips off. For this reason the underside of the stone between the groove and the edge is called a drip. A detail of this is shown in Fig. 249.

Cast stone copings: Because natural stone is expensive, cast stone copings are commonly used today. Cast stone is made by specialist firms. The stones are

made with a core of concrete faced with a mixture of crushed stone particles and cement. The surface of cast stone soon shows irregular unsightly staining.

D.p.c. beneath the coping stones: Coping stones are usually in lengths of 600 and the joints between them are filled with cement mortar. In time the mortar between the joints may crack and rainwater may penetrate and saturate the parapet wall below. If frost occurs the parapet wall may be damaged. To prevent the possibility of rainwater saturating the parapet through the cracks in coping stones it is common practice to build in a continuous d.p.c. of

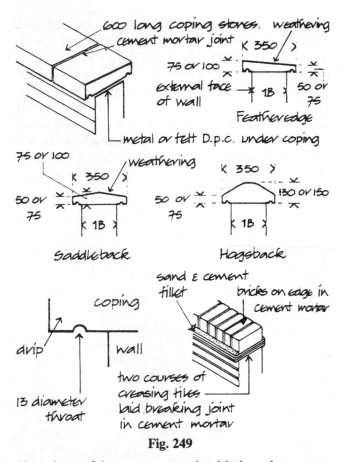

Fig. 249

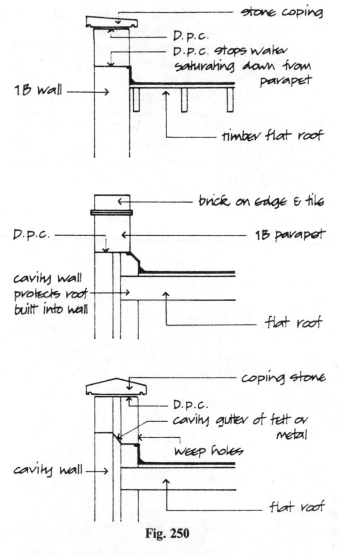

Fig. 250

bituminous felt or copper or lead below the stones. This is illustrated in Fig. 249.

Brick cappings: Another method of capping parapet walls is to form a brick on edge and tile creasing capping. This consists of a top course of bricks laid on edge, and two courses of clay creasing tiles laid breaking joint in cement mortar as shown in Fig. 249. The bricks of the capping are laid on edge, rather than on bed, because many facing bricks have sand faced stretcher and header faces. By laying the bricks on edge only the sanded faces show whereas if the bricks were laid on bed, the bed face which is not sanded would show. Also a brick on edge capping looks better than one laid on bed. Creasing tiles are made of burned clay and are usually 265 long by 165 wide and 10 thick. The tiles are laid in two courses breaking joint.

It will be seen that the tiles overhang the wall by 25 to throw water away from the parapet below. A weathered fillet of cement and sand is formed on top of the projecting tile edges to assist in throwing water away from the wall. Two courses of good creasing tiles are sufficient to prevent water soaking down into the wall and no d.p.c. is necessary under them.

Parapet walls should always be built with sound, hard, well burned bricks which are much less liable to frost than common bricks. The bricks should be laid in cement mortar, mix 1 cement to 3 of sand.

Parapet wall d.p.c.: It is good practice to build a continuous horizontal d.p.c. into brick parapet walls at the junction of the roof covering, upstand or skirting with the wall. Bitumen felt and asphalt covering a flat roof is turned up against the parapet as a skirting 150 high. It is at this level that a d.p.c. is usually built as shown in Fig. 250. This d.p.c. may be of bituminous felt, lead cored felt or copper or lead sheet. A metal d.p.c. may be continued as a weathering over the upstand of bitumen or asphalt as a flexible weathering to accommodate slight relative movements between roof and wall and in addition provide ventilation to timber flat roofs.

Parapet to cavity wall: The construction of a parapet built on a cavity wall is usually somewhat different from that built on a solid wall. An external wall is built with a cavity to prevent rain penetrating the wall and it is logical to continue the cavity to at least the top of the roof, so that the cavity protects roof timber or concrete built into or against the wall. This construction is illustrated in Fig. 250. It will be seen that the cavity is continued to the level of the asphalt skirting. Above the d.p.c. the parapet is built in solid 1*B* work.

An alternative construction sometimes used is to continue the cavity wall up to the coping or capping as shown in Fig. 250. No good purpose is served in continuing the cavity walling up in the parapet as the cavity above the roof no longer acts as a barrier to penetration of moisture. The cavity is moreover an inconvenience in the parapet as moisture may penetrate the brick skins and water collecting in the cavity has to be drained out by the continuous cavity gutter shown.

INDEX